Bioethics in **Perspective**

Corporate power, public health and political economy

In *Bioethics in Perspective* Scott Mann demonstrates the importance of issues of corporate power, global inequality and sustainability in shaping health outcomes around the world. The text develops a comprehensive ethical and practical critique of the neoliberal economic ideas which have guided policy in the English-speaking world. It explores the consequences of such policies for health and health care around the world, in terms of increasing health inequalities, serious food and water shortages, inadequate health care provision and the marketing of dangerous and unnecessary drugs.

With clear proposals for political and economic reform to effectively address these problems, *Bioethics in Perspective* provides an important counterbalance to much conventional commentary on bioethics. It takes readers with little or no prior knowledge of ethics, economics or medicine quickly and easily into advanced debates and discussions about the causes and consequences of health and illness around the world.

Scott Mann is Associate Professor in the School of Law at the University of Western Sydney.

Bioethics in Perspective

Corporate power, public health and political economy

Scott Mann

CAMBRIDGE UNIVERSITY PRESS
Cambridge, New York, Melbourne, Madrid, Cape Town, Singapore,
São Paulo, Delhi, Dubai, Tokyo, Mexico City

Cambridge University Press
477 Williamstown Road, Port Melbourne, VIC 3207, Australia

Published in the United States of America by Cambridge University Press, New York

www.cambridge.org
Information on this title: www.cambridge.org/9780521756563

© Scott Mann 2010

This publication is copyright. Subject to statutory exception
and to the provisions of relevant collective licensing agreements,
no reproduction of any part may take place without the written
permission of Cambridge University Press.

First published 2010

Cover design by David Thomas
Typeset by Aptara Corp.
Printed in China by C&C Offset

A catalogue record for this publication is available from the British Library

National Library of Australia Cataloguing in Publication data
   Mann, Scott.
   Bioethics in perspective : corporate power, public health and political
   economy / Scott Mann.
   9780521756563 (pbk.)
   Includes index.
   Bioethics.
   Medical ethics.
   Medical policy.
174.957

ISBN 978-0-521-75656-3 paperback

Reproduction and communication for educational purposes
The Australian *Copyright Act 1968* (the Act) allows a maximum of
one chapter or 10% of the pages of this work, whichever is the greater,
to be reproduced and/or communicated by any educational institution
for its educational purposes provided that the educational institution
(or the body that administers it) has given a remuneration notice to
Copyright Agency Limited (CAL) under the Act.

For details of the CAL licence for educational institutions contact:

Copyright Agency Limited
Level 15, 233 Castlereagh Street
Sydney NSW 2000
Telephone: (02) 9394 7600
Facsimile: (02) 9394 7601
E-mail: info@copyright.com.au

Cambridge University Press has no responsibility for the persistence or
accuracy of URLs for external or third-party internet websites referred to
in this publication and does not guarantee that any content on such
websites is, or will remain, accurate or appropriate.

Contents

Introduction	1
1 Ethics and ideology	10
2 Freely competitive markets	36
3 Problems of competition	60
4 Health inequalities	96
5 Food and water	117
6 GM foods and life patents	146
7 Energy and the greenhouse effect	170
8 Health care provision	198
9 Drugs, drug testing and drug companies	221
10 What is to be done?	256
Bibliography	269
Index	274

Introduction

Biomedical ethics

This book provides an alternative perspective on bioethics. It argues the case for making issues of corporate power, global inequality and sustainability central elements of the syllabus, insofar as they are fundamental in shaping health outcomes around the world, today and in the future.

The argument is that the neoliberal ideology, which has dominated economic and political thinking and decision making in much of the world for the last 30 years, has had generally dire consequences for the health of people, particularly poorer people, and of the planet. The book shows how and why this has been the case, and what needs to be done to improve the situation.

Most research projects, tertiary units and texts in bioethics focus upon ethical and legal issues of professional responsibilities of doctors, nurses and biomedical researchers. In particular they look at the responsibilities of such professionals to respect and protect the autonomy, rights and well-being of patients and research subjects including animal subjects of biomedical research.

Such projects, units and texts consider issues of justice and fairness in relation to provision of health care, the ethical implications of research into, and application of, new biomedical technologies of gene transfer and genetic diagnosis, stem cell manipulation and cloning, in-vitro fertilisation and other new assisted reproductive technologies. They examine a range of traditional ethical and legal 'problem areas', including abortion, euthanasia and involuntary civil commitment.

Such ethical consideration has guided professional practice. It has influenced law reform and the regulation of research, and has encouraged public discussion and debate. However, focusing largely or completely upon these particular areas can foster an impression that these are the only, or the most significant, ethical issues associated with medical care and medical research;

that issues of life and death are all and only about access to life-saving treatments, abortion, euthanasia and the risks to medical research participants.

Most of these issues are actually what could be called 'late stage' issues. They concern the later stages of often complex sequences or causal chains of social developments at a global level, including broad social policy decisions and applications, bearing upon the health of individuals and populations. Interventions at this relatively late stage of the causal chain are generally not capable of addressing the major structural factors that shape the health experiences of populations. The evolution of bioethics as a discipline has contributed to the failure to address such significant structural issues by encouraging a belief that it is only end stage issues of the conduct of medical treatment and medical research that are the significant or appropriate content for ethical consideration.

The consequences of too narrow a focus upon end stage issues can be seen in many areas of 'mainstream' bioethics. In relation to research ethics, for example, it has contributed to a situation where 'ethics review' now involves consideration of a very narrow range of 'specifically ethical' issues, increasingly segregated from methodological concerns, and from any broader consideration of the social context of the research in question. This means that research proposals can get 'ethics approval', and thus claim that they are indeed 'ethical' by ticking a series of boxes relating to confidentiality, consent, risk and recruitment, when, in fact, such proposals could be seen to be highly unethical in their basic assumptions, aims, direction, opportunity costs and likely consequences, when viewed in a broader social context.

Another example is provided by the increasing concern of bioethicists with 'ageing populations' and the burgeoning costs of new life preserving and life extending treatments. They debate issues of who should get such treatments, whether older people should be sacrificed for younger and whether anything approaching comprehensive public provision, where it still exists in the developed world, can possibly be sustained in light of these developments.

Such debates typically fail to make any reference to the trillions of dollars spent by the US and its allies in the massive destruction of people and property in Iraq, to the failure of the US Government to use its market power to bid down outrageous monopoly drug prices and its efforts to stop other governments from doing so, and to the success of Cuba in continuing to provide comprehensive public health care despite its relative poverty. Failure to consider these issues renders such discussions themselves deeply un-ethical.

In order to make properly informed ethical sense of such important end stage issues of the conduct of research projects and provision of medical care, it is crucial to see them in their broader social–structural context. This book will assist in this process by providing a complement to the orthodox system of priorities and the orthodox syllabus.

Neoliberal ideas and policies

The primary focus of this book is, therefore, upon some major social-structural determinants of good and bad health in the world today, and in the immediate future. It seeks to identify those social structures, relations and forces and those material technologies which are most important in shaping patterns of good and bad health around the world. In doing so, it focuses particularly upon the role of market forces and corporate power in generating inequality and poverty, and shaping provision or deprivation of health care, food, water, energy and medicines.

The basic thesis is that, while there have been significant improvements in some areas, including increasing average lifespan, the neoliberal ideas and policies which have shaped economic and political developments around the world for the last three decades have had, and continue to have, disastrous and worsening effects upon human health and well-being. These improvements have been achieved in spite of, rather than as a consequence of, such ideas and policies. A great deal can – and needs – to be done in order to dismantle and replace key ideas, institutions, practices and relations created or maintained throughout this period.

In particular, neoliberal ideas and policies have contributed to increasing inequality within and between countries, and increasing poverty and deprivation for those at the bottom of the heap. Such increasing inequality and poverty have contributed to much worse health outcomes than could have been achieved through policies designed to reduce such inequality.

Neoliberal ideas and policies have contributed to accelerated pollution and depletion of fossil fuels and accelerated global warming through unrestricted pursuit of profit. The principal perpetrators, the big corporations who mine and burn coal, who pump oil out of the ground and sell cars and trucks to burn such oil, who continue to impose oil based industrial agriculture on the world, also use their market power to prevent any effective political action to sheet home to them the real costs of their operations – or to move to sustainable low emission, low toxicity systems before the climate and the soil are further irreversibly damaged.

Such damage, including reduced water supplies, increased aridity and land degradation in key regions, and such misguided responses to peak oil and global warming as the use of food crops to produce bio-fuels (leading to increased food costs), is already pushing millions more poor people each year into food and water insecurity, with all the devastating health consequences of such insecurity. Moreover, the process will rapidly accelerate without urgent and far reaching corrective action.

Issues of moral responsibility are explored in this context, by consideration of the role of different interests, groups, and individuals, within the social structures and processes most centrally involved in shaping and determining patterns of good and ill health around the world. This certainly involves reference to different systems of provision of medical care in different nations and different regions.

But the book also looks at the organisation of production and distribution – by corporations, governments, regulatory authorities, international agencies and market forces – of food, water and energy, as the foundations of human life and human health on the planet. It focuses on the production and distribution of drugs as an increasingly significant and costly part of medical care and treatment around the world.

As with most bioethics texts, the book begins by considering the general nature of ethical theories and ethical concepts, prior to considering the application of such ideas in the biomedical area. It also argues for the necessity for considering key elements of economic and political theory, and the current state of the world economy, shaped and influenced by such theory, in order to properly appreciate the significance of such moral ideas for the production of health and illness in the modern world.

It then provides the necessary background of economic theory and economic analysis to enable deeper consideration of conflicting ideas or ideologies of rights, justice and social welfare in contemporary society. Such economic theory allows for understanding of the central role of economic forces in shaping and determining the health and ill health of people and nations. Building upon this foundation, it considers some of the ways in which neoliberal ideology and policy have impacted most directly upon the health of populations, groups and individuals, through the provision, or failure of provision, of healthy food and clean water, energy, medical care and pharmaceutical products. It also looks at issues of the corporate domination of medical research involving human and animal subjects.

Science

This book is critical of a number of contemporary developments associated with the application of science to production and pollution control, including the industrialisation of agriculture, the spread of genetically engineered crops, nuclear power and carbon sequestration. It also criticises a number of the actions of big drug companies, frequently justified by reference to science and scientific method. However, it is in no way opposed to science per se or to the application of scientific research in the development of new technologies.

I have tried to ensure that all of the ideas and arguments of this book, including its criticisms of neoliberal economic theory and policy, of industrial agriculture, genetically engineered (GE) crops and other technologies, are built upon a basis of coherent, well-supported natural and social scientific theory and relevant empirical data. I believe that science, appropriately guided by ethics and politics, is the only solid foundation for understanding, and for acting to effectively address, the major issues and problems under consideration in this book.

In too many cases, supporters of 'new' technologies (e.g. of herbicide resistant crops) appeal to the role of scientific research in the development of such technologies to try to justify their continued or expanded application, when, in fact, the empirical evidence shows such technologies to be destructive, dangerous, unnecessary or unnecessarily costly, compared to readily available alternatives.

Supporters of some existing technologies (e.g. of coal fired power stations) argue for the necessity for developing radically new technologies (e.g. of carbon sequestration) in order to address major social and environmental problems. They try to brand critics as anti-scientific, when, in fact, the empirical evidence shows that there are major problems with both the existing and the proposed future technologies, and that appropriate development and application of other existing technologies (e.g. of wind and solar power) – or changed social and political relations – can effectively address the problems in question and radically transform the world for the better.

There is nothing anti-scientific about highlighting the relevant evidence and arguments. This does not, in any way, imply any hostility to further scientific research or to the practical application of such research. However, it does imply appropriate selection of the areas where such research can be most effectively and responsibly applied.

Amongst other things, science essentially concerns the submission of truth claims to relevant empirical testing. For the purposes of this book the crucial issues are always those of the truth or falsity of claims made and the strength of relevant evidence supporting or refuting such claims. This includes claims about theories (of society and of nature) and about facts, including claims about the efficacy or safety of, or necessity for, particular technologies.

Summary of chapters

Chapter 1 looks at ethical theories and concepts, highlighting some problems with key ideas of deontology and utilitarianism. It considers the ways in which 'real world' ethical decision making is shaped by the dominant ideologies within which ethical ideas are embedded. It demonstrates the need to ground ethical ideas in a coherent economic and political framework. It draws upon key ideas and analyses of moral philosopher Richard Norman.

Chapter 2 introduces the neoliberal model of economic development based upon unregulated markets and corporations, and related ethical ideas of just deserts, which have dominated ideology and policy, particularly in the

English-speaking world, for nearly thirty years. This chapter draws, particularly, upon the work of Ha-Joon Chang and Ilene Grabel.

Chapter 3 provides a systematic critique of the neoliberal economic ideas by reference both to basic issues of the logical coherence of such ideas and to the real facts of development of the world market in the period of neoliberal domination, including the production of the current recession. The chapter shows the factual, practical and ethical bankruptcy of the theory, and in so doing, points the way towards possible alternative approaches. This chapter uses ideas drawn from the work of Ha-Joon Chang, Paul Mason, George Soros, Harry Shutt and Joseph Stiglitz.

Chapter 4 provides a concise overview of broad patterns of health and ill health around the modern world, as well as some historical background and the beginnings of some explanations for significant health inequalities. It draws particularly upon the work of Hilary Graham, Shereen Usdin, Michael Marmot, Richard Wilkinson and Bruce Kennedy in exploring the connections between health inequalities and other social inequalities.

Chapter 5 casts further light upon such patterns and inequalities by reference to the production and distribution of food and water. It explains how and why significant numbers of people are currently 'food insecure' and why such numbers will inevitably increase in the future with disastrous health consequences, unless and until there are major policy changes. Here, the investigation has been guided by the work of Vandana Shiva, Walden Bello, Tony Weis and Paul Roberts on the present and likely future state of world agricultural production.

Chapter 6 continues the investigation of the current and future state of world food supply by considering the promise and problems of genetically modified plants and animals as major food sources today and in the future. It focuses, in particular, upon health risks associated with the current generation of genetically modified food plants. It briefly addresses some of the issues and problems of life patents. In this case, Antoinette Rouvroy's investigation of genetics and neoliberalism, Sheldon Krimsky's analysis of gene patents and Jeffrey Smith's detailed documentation of currently available evidence of the dangers of GM crops have been the principal guides.

Chapter 7 focuses on energy provision as the other major pillar of material support for contemporary economic life and health. It looks at the role of fossil fuel combustion in producing global warming, and at some of the likely health consequences of such warming. It considers the impending energy crisis as world oil supplies wind down, and its implications for health and the various different proposals and possibilities for addressing both global warming and peak oil. The work of Mark Diesendorf, Mark Maslin and Helen Caldicott has provided help and guidance here.

Chapter 8 looks at health care provision around the world. It compares the neoliberal ideology of private health provision with the reality, and it compares predominantly private with predominantly public systems. The writings of Lawrence O. Gostin and his collaborators in the area of public health law have provided valuable assistance in this area.

Chapter 9 looks at the development and marketing of drugs by the big pharmaceutical corporations, and the role of the US Food and Drug Administration (FDA) in regulating – or failing to regulate – these processes. It briefly considers the traditionally central bioethical problem areas of human and animal research. The chapter looks at the role of Big Pharma in shaping intellectual property protection in international law and in bilateral trade agreements. It looks at the US–Australian Free Trade Agreement and its likely consequences for the cost and availability of prescription drugs in Australia. Lastly, it looks at the use of the vulnerable developing world populations as subjects of drug trials by developed world corporations. The works of Marcia Angell, Sonia Shah, Ray Moynihan, Andrew Cassels, Linda Weiss, Elizabeth Thurborn, John Matthews and Sheldon Krimsky have inspired and informed the discussion of these issues.

Chapter 10 begins to explore some possible solutions to the problems identified in earlier chapters. It highlights the necessity for rational and democratically regulated planning and redistribution to replace the operation of unregulated market forces and undemocratic corporate power in order to effectively combat inequality and poverty and significantly improve health outcomes for currently disadvantaged groups. Such planning is crucial for creating genuinely sustainable technologies in order to maintain and extend such health gains into the future.

It will be clear that, in addition to the core issues of the traditional bioethics syllabus touched upon earlier, a number of other key issues of public health and ill health are not covered. In particular, the book does not directly deal with the dire public health consequences of warfare, the crucial role of education, or the lack of it, in shaping public health outcomes, or the role of civil liability legislation, tort law, mediation and no fault schemes in protecting – or failing to protect – the health of individuals and populations.

Failure to directly address these issues in no way reflects any judgement about their lack of importance in shaping public health. On the contrary, it is rather the depth and complexity of the issues involved that precludes adequate coverage of these issues on this occasion.

Reforms

Most of the issues discussed in this book involve major problems in contemporary political, legal and economic arrangements. They concern practices and institutions that urgently need to be ended, dismantled, changed, replaced, compensated for, or repaired. This is by no means intended to suggest that all is lost or that there are not plenty of positive developments. The book highlights at least some of these positive developments at various points in the discussion, particularly where they can be built upon to produce further progress in the future.

The book does not spend much time attributing blame for the problems it identifies. The major problem lies in the fact that we are all trapped within a social and economic system, built around competition, widespread disempowerment and fear. We are all struggling to survive, to try to protect ourselves and our families, and, apparently, forced to act to maintain that system in order to do so or to have any kind of worthwhile life.

While elements of the previously prevailing neoliberal consensus have been thoroughly discredited by the current world economic crisis, nonetheless, the key underlying theme of 'objective market forces beyond human control' remains just as strong.

Indeed, with an increasing number of people starving, and a massive future food crisis looming, with water and oil supplies running out and global warming accelerating, with AIDS still killing hundreds of thousands of people every year, mainly in sub-Saharan Africa, world leaders debate about how they can put a massively corrupt, immoral, inefficient and broken financial system back together again with the minimum of change or threat to the wealth and luxurious lifestyle of the power elite. Apparently 'objective forces' make any other kind of system impossible.

Too many of those with the power to directly shape future events seem to have little plan for the future beyond the immediate preservation of their own power and privilege. Half-hearted policies of 'Corporate Social Responsibility', 'governance reform' and 'carbon trading' provide a cover for continued 'business as usual', including the accelerated dismantling of democratic rights, the 'poisoning of the biosphere' and the 'deterioration in the conditions for human life' (Kempf, 2009, p. 59), rather than driving any real social change or social progress.

At the same time, it is important to see that many of the processes and problems, identified by currently ruling ideologies as beyond all possibility of human control, are, in fact, quite amenable to being effectively addressed by concerted human action.

There is increasing recognition amongst ordinary citizens that our political leaders could and should be dismantling 'free markets' in key areas of the world economy. They should take the banks and big corporations into democratic, public ownership, to direct their resources by rational planning, to repair and regenerate a world that raw capitalism has almost destroyed. In order to do this, they need to mobilise the mass of the population, rather than follow the dictates of the rich and powerful beneficiaries of the current system. If they fail to do this, the mass of the population needs to take the initiative in directing such processes.

The current economic crisis offers real opportunities for positive, practical and ethical change; for repairing the social, psychological and ecological damage inflicted by years of unfettered greed, cruelty and stupidity, justified and driven forward by 'economic rationalism'.

The danger, and the likelihood, is that such opportunities will be missed; the system will be temporarily patched up to allow continued 'business as usual' in the developed world for a few more years, ignoring the chronic and worsening crisis in the rest of the world and paving the way for a yet more devastating world economic crisis a few years down the track.

This book aims to clarify some of the ways in which prevailing economic and political ideas and policies have impacted upon major health issues and problems, with a view to encouraging new and creative thinking about possible improvements and solutions.

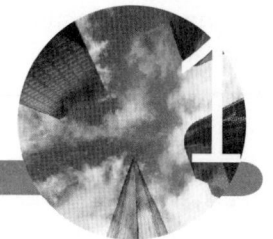

Ethics and ideology

This chapter introduces some major ethical theories and principles and explores the relationship between ethical ideas and dominant belief systems or ideologies. It begins to show how currently dominant ideologies shape ethical thinking and decision making.

Ethics

Ethics is a branch of philosophy concerned with serious consideration, clarification and rational assessment of moral ideas and moral decision making. This inevitably involves abstraction and isolation of moral ideas from their 'real world' contexts, where they exist as elements of complex systems of belief and forms of life, along with idealisation and logical development of such ideas. While this helps in understanding the ethical dimension of all ideologies, and encouraging individual recognition and reassessment of previously taken-for-granted moral ideas, it can also function to obscure the details of specific belief systems or ideologies which actually shape real-world moral decision making.

Ethicists have sought to clarify the key features of specifically moral thought and action. In particular, they have distinguished moral judgments of the rightness or wrongness of particular actions from legal, aesthetic or prudential judgments. Rather than being concerned with principles of law, of taste, or of individual well-being, moral judgments depend upon specifically moral standards – of value, virtue and obligation, of ultimate or intrinsic worth, of what we ought, or should, or are obliged to do, without condition; of the rights and well-being of others, ultimately of all humans or of all sentient life, rather than pure self-interest.

Ethicists have frequently argued that morality is grounded in empathy – the individual's capacity for putting themselves in others' shoes. This enables people both to gain insight into how others would like to be treated and how they themselves would like to be treated if they were in the other's position. Empathy with particular individuals validates and guides the formulation of general principles applicable to all individuals. On the other hand, ethical principles can also counteract individual empathy through placing higher goals (e.g. justice, freedom) above the suffering of individuals.

Philosophical consideration of moral decision making has identified two broad categories of ethical theory seen as guiding and directing such decision making: teleological theories and deontological theories. Teleological theories view judgments of obligation as grounded in judgments of value (Waluchow, 2003, p. 35). Having established what is of genuine ethical value, we are obliged to try to achieve or promote that thing. The most influential and significant teleological theories are utilitarian theories. These identify human happiness or well-being as the only true value or intrinsic good, and therefore command that we do whatever will create more human happiness or well-being than anything else.

Deontology

Deontological theories, by contrast, focus upon the nature, rather than the consequences, of actions as of major ethical significance. In this connection, Immanuel Kant refers to 'duty' performed for its own sake, rather than 'in order to promote human happiness or fulfilment' (Norman, 1983, p. 95, p. 97). He maintains that 'nothing is unconditionally good except a good will'. Moral evaluation should focus upon the intention with which an action is performed, rather than its consequences.

Deontological theories see certain sorts of action as morally good – and therefore more or less mandatory – in themselves, rather than by virtue of their contribution to achieving some other good. Acting with good intentions, obeying the Ten Commandments, following the golden rule (requiring individuals to try to treat others as they would want to be treated if they were in the same situation) or respecting and promoting the autonomy and rights of other people, are actions widely recognised as intrinsically, morally good. Here we can distinguish issues of duties to respect the universal human rights of all others, as opposed to duties to respect and honour specific promises and commitments made to specific others.

Rights are legitimate entitlements of individuals to particular goods and to the free exercise of particular abilities and powers. Rights imply responsibilities of others, not to infringe upon particular individuals' rights and to respect or facilitate the effective realisation of the rights in question (Nickel, 2007).

Legislation and judge-made law create legal rights as entitlements given the force of legal sanction. For example, contract law empowers those who have accepted the promise of another to provide some good or service and have themselves contributed some consideration to the promiser, without receiving the good or service in question. It enforces their right to compensation for non-performance or to 'specific performance' of the contract in question; where the court orders the defaulting party to carry out their contractual obligations.

But 'human' rights are rights that all people have at all times by virtue of being human. If such rights are not respected then people cannot live properly human lives. Many would see health care as occupying a key role here. If seriously ill people are denied access to health care they cannot live a good life, quite likely they cannot live any life at all.

Here we can distinguish what might be called welfare rights on the one hand and autonomy rights on the other. Welfare rights can be seen as grounded upon human needs, upon the objective requirements for human survival, health, fulfilment and happiness, and the realisation of characteristically human potentials. If people are going to survive and prosper then they must be assured the necessary means for doing so with food, water (for drinking, washing and waste disposal), shelter, education, love, support, and recognition, as well as health care. If they are going to be assured of such access then they must be provided with the necessary means for achieving such 'goods' in the particular society concerned. In a capitalist society, people need money – or welfare state provision – to get access to basic material necessities. They need a job or another source of income in order to get access to money. These things need to be considered human rights too.

Autonomy rights, by contrast, focus upon individuals' capacities to make their own life choices and act upon such choices. This means providing people with the mental, social and physical tools to be able to make rational and informed choices, and act upon them. This includes rights to high quality education and access to reliable information about the world. Democratic rights – to free speech and association, to voting and standing for political office, and legal rights to a fair and speedy hearing in a court of law if detained by the justice system – can be seen as expanding the range of individual autonomy by empowering individuals to exercise some control over political and legal decision making. At the same time, there is evidence that individuals suffer psychologically and physically if they do not feel in control of their own lives; there is a human need for autonomy.

Laws can be enacted that aim to ensure that human rights are respected by making them legal rights. The United Nations' *Universal Declaration of Human Rights* (UDHR) of 1948 sets out a range of basic human rights, including both welfare rights (of access to objects of need and to the means of access to such objects) and autonomy rights, including political and legal rights, which should be equally available to all people at birth 'without discrimination'. Subsequent human rights treaties oblige their signatories to ensure the protection of all such rights through their own national legal systems.

Particularly important for present purposes are the economic, social and cultural rights set out in Articles 22–27 of the UDHR, including the right to work, with equal pay for equal work; to join trade unions; to reasonable working hours and holidays; to an adequate standard of living (including food, clothing, housing and medical care); to benefits in the event of unemployment, sickness, disability or old age; and to free education. As Geoffrey Robertson (2009, p. 37) says, the contemporariness of this document still 'amazes', and it has inspired more than 200 – subsequent – international treaties, conventions and declarations and the bills of rights found in many countries.

The *International Covenant on Social, Economic and Cultural Rights (1966)*, which makes such rights a foundation of international law, states (in Article 12) that the signatories 'recognize the right of everyone to the enjoyment of the highest attainable standard of physical and mental health' including

> provision for the reduction of the stillbirth rate, and of infant mortality and for the healthy development of the child; improvement in all aspects of environmental and industrial hygiene; prevention, treatment and control of epidemic, endemic, occupational and other diseases; [and] creation of the conditions which would assure to all medical service and medical attention in the event of sickness. (Flynn, 2003, p. 63)

Humans owe respect and support for the human rights of all other humans because they are humans. But so too do people choose to create or maintain social relationships with others which carry with them specific moral responsibilities. By making commitments or promises, they incur specific moral responsibilities to act in the expected fashion and give others the moral right to expect such action from them. This includes commitments associated with family, friendships and work relations, such as to assist and care for others, not to harm them, to be truthful with them, to prioritise their interests and their welfare over that of others, to try to compensate them for wrongs inevitably done to them.

The sort of contractual arrangements dealt with by contract law may or may not be seen to have such ethical significance. Such arrangements may, for example, be essentially 'unconscionable' in legal terms, in the sense of resulting from and reflecting a radical imbalance of knowledge and power, rather than free, rational and informed decision by both parties.

Utilitarianism

The basic idea of utilitarianism is for an individual to try to maximise the happiness or well-being of all those affected by that individual's actions. In any situation, the requirement is for an individual – or group – to consider all the actions available to them and all the people who will be affected by such actions. They then have to pick the action with the greatest balance of utility or social welfare over disutility – or the smallest balance of disutility over utility – for all concerned. No one's particular interests or welfare should be prioritised over anyone else's, but a minority might have to lose out in order to further the well-being of the majority (Norman, 1983, Chapter 7; Waluchow, 2003, Chapter 6).

Ethicists have highlighted numerous practical and theoretical problems with utilitarianism. It seems particularly difficult for an individual to apply utilitarian principles in their daily life because such principles are likely to conflict with special commitments to family, friends and colleagues. It is difficult to map out all the future welfare consequences of all possible choices in the present. It is difficult to provide objective, quantitative measures of the kind required by the utilitarian calculation in relation to such a subjective thing as happiness. On the other hand, well-being lends itself much more readily to such quantification via such measures as mortality and morbidity, including infant mortality, and levels of income, stress, crime and education within a region and within different social class groups. Such information can, at least, guide utilitarian decision making by public authorities and by citizens involved in political practice (Wilkinson and Pickett, 2009).

Here again, we see the importance of empirical facts about good and bad health in putting ethical theory and ethical judgment upon a genuinely objective foundation. In this case, the World Health Organization (WHO) country health indicators of life expectancy, literacy rates, disease prevalence and gross national product (GNP) per capita, provide a first step towards a utilitarian assessment of social policy decision making affecting the countries in question (including relevant policy decisions in other countries).

Assuming the possibility of effective quantification (of well-being), there are problems of unfairness associated with simple maximisation. Average life expectancy in a region could and does mask radical disparities of 5, 10, 15 and 20 years between different class groups; GNP per capita could and does mask radical disparities of income between such groups. Most people's ethical intuitions – about the nature of justice and fairness – support some loss of overall utility in favour of a more even division of available utility.

Responses to this issue of the apparent unfairness of utilitarianism have included support for maximin principles: choosing the option that maximises the utility of the person receiving the minimum amount; Pareto principles which endorse changes that make no one worse off; or for going for the highest level of utility universally achievable for each person concerned. The influential contribution of political philosopher John Rawls, seeking to address this issue of justice and fairness, is considered later in this chapter.

There are issues and problems with all such proposals. As has been frequently pointed out, some idea of a minimum of basic human rights for everyone concerned effectively counteracts negative quantification features of utilitarianism. But the major benefit of utilitarianism is precisely its capacity to come up with a solution in situations of apparently otherwise irreconcilable conflicts of rights.

The sorts of examples presented in ethics textbooks tend to highlight this conflict-of-rights-resolving role by focusing upon immediate life and death decisions. The fact that a decision has to be made about whether to shoot down the hijacked aircraft over the countryside before it has crashed on the urban centre or nuclear waste storage facility seems to show that it is impossible for responsible rulers not to be utilitarians, at least some of the time. The latter sort of life and death decision does, indeed, inevitably involve a complete sacrifice of the rights and welfare of a minority.

Most real-world political decisions do not involve such immediate and immediately devastating consequences. But a social policy decision that operates in such a way as to channel limited resources to one particular group and away from another, such that the members of the first group live, on average, 15 years longer than the members of the second group, is also a life and death decision. The shorter-lived group have lost the extra years of life a more even division of resources could have given them.

While maximisation seems, at first, a potentially deeply unfair principle, there is actually quite a lot of empirical evidence which suggests that, if properly applied, it is more likely to push social policy decision making towards greater equality. Smaller improvements in mortality and morbidity figures for the majority are likely to boost averages more than bigger gains for a privileged minority. As will be shown in later chapters, reduced inequality can create huge gains for the poorer majority without significantly reducing real welfare measures for the more privileged.

The rapidly declining marginal utility of wealth, on any reasonable measure of real social welfare benefit, means that serious utilitarian calculation calls for immediate and deep-going redistribution of existing wealth to ensure the satisfaction of the basic needs of all. There can be no utilitarian justification for anyone having two or 20 houses when others have none, or someone having much more food than they need while someone else starves.

This does not prevent supporters of radical inequality seeking to provide utilitarian justifications of such inequality. In particular, it is often claimed that significant inequality can produce much higher overall output of goods than greater equality, such that, for example, either the majority or the worst off are actually significantly better off with greater inequality. This is, of course, a factual, rather than a value claim. It immediately raises crucial questions of economic analysis and prediction.

Animals

The originator of utilitarian theory, Jeremy Bentham, recognised that insofar as animals, as well as people, experience pleasure and pain and can be seen to have their own interests in minimisation of pain and harm then their interests need to be included in any genuine utilitarian calculation. Also, because they can be subject to just as much suffering as human beings, they have a right to equal consideration along with humans in any such calculation (Kuhse and Singer, 2006).

Considerations of the nature of animal consciousness, of the existence of moral sentiments and behaviours in many types of animals, and of issues of animal welfare and animal rights are areas of strength in contemporary bioethics as a result, in particular, of the work of Peter Singer, who has pushed these issues to the forefront of contemporary consideration.

While these issues have received increasing consideration in theory, in practice the situation is not good, with continued abuse of animal life on a vast scale, particularly through habitat destruction and industrial farming, but also through cruel and inhumane conditions and practices in hunting, zoos, circuses and laboratories. Issues of animal abuse in factory farming and in drug research are touched upon in Chapters 5 and 9.

Just deserts

Utilitarianism seeks to distribute costs and benefits in such a way as to maximise overall social welfare, and this could well imply a more equal distribution. Deontology calls for equal respect for the human rights of all. However, many people would argue that a fair or just distribution should take account of what particular individuals and groups actually deserve. We should, in other words, consider a utilitarian or deontological requirement for equality as abstracting from individual differences in this regard.

Some maintain that once we take account of such differences, then we can no longer justify equal distribution of welfare benefits or even equal respect for human rights. Some deserve more good things by virtue of their greater social contribution; some deserve less, by virtue of their lesser contribution. Indeed, some can expect a lesser regard for their human rights by virtue of their 'negative' social contribution in the form of criminal activity. They might, indeed, even lose significant rights to property, liberty or life itself if they choose to radically disrespect the human rights of others.

Others argue strongly for the unqualified protection of basic rights of all people, including those judged to have broken the law, who should continue to be treated with appropriate care and respect for both their welfare rights and legal and political rights.

It is generally agreed that for the principle of just deserts to make sense there must be some approximate 'equality of opportunity' for social contribution. It would not be fair to reward some who choose to make a particular contribution when others are deprived of the opportunity to make such a choice. At least in situations of pressing need for relatively scarce talents, equality of opportunity can also be justified in utilitarian terms as allowing optimal creative and productive development of human resources – no one is denied the opportunity for such development.

A common idea here is that what people should get out of overall social production should be a function of what they put in; that those who put in more – thereby benefitting others more – should legitimately expect greater rewards. As considered in later chapters, there are different ideas of what sort of social contribution should form the basis of such differential reward. Some argue that individuals should be rewarded in proportion to the actual value of the social wealth they create. Others say that inevitable inequality of opportunity will mean that some people will always be in a position to create more wealth than others through the same or a lesser level of effort or sacrifice, and it is therefore fairer that individuals be rewarded in proportion to the socially valuable effort, sacrifice or risk they put in, rather than the actual value of their output.

Some reject just deserts altogether in favour of distribution of goods according to need. This raises difficult practical issues of how to motivate intrinsically unrewarding but socially valuable or necessary labour without appropriate rewards. It is theoretically possible to combine just deserts and distribution according to need through combining tax funded welfare provision with reward for contribution in producing goods for sale in a market place.

Welfare rights have already been closely associated with an idea of human needs, or things 'without which people cannot live properly human lives'. If we clearly distinguish goods which are genuinely needed by all people, including 'basic food and clothing, a house to live in, medical care' and education, as 'necessary prerequisites for any worthwhile human life' (Norman, 1987, p. 118) from other things which people would merely like to have, then we can call for the provision of the former sort of necessary goods free to all by public authority, with other goods 'purchased by individuals according to their individual preferences'.

Such public provision has traditionally been paid for through various forms of taxation, or through the profits of public enterprise. This raises crucial questions of what constitute fair or just systems of taxation, or of how the necessary labour will be fairly allocated in the absence of such taxation.

Social, personal and professional ethics

Ethicists have drawn a distinction between social morality on the one hand, and personal and professional morality on the other. Social morality is concerned with how society should be organised, what laws should be passed and enforced. Personal and professional morality is concerned with how individuals should act in relation to others with whom they have particular family or work relations. Bioethics has developed as a mix of the two; on the one hand there is a principal focus on ethical issues of professional responsibility – of doctors, nurses and medical researchers. On the other, there is some consideration of policy issues in relation to health care provision and the direction of research.

The differentiation of social ethics on the one side and personal and professional ethics on the other can suggest that personal and professional morality has nothing to do with social policy issues, and that social policy issues are purely the preserve of politicians, senior civil servants and high court judges – and perhaps also chief executive officers (CEOs) of big corporations. Social policy issues are, in fact, deeply intertwined with issues of professional and familial responsibility. In a politically relatively open, at least partially democratic, society all citizens have some possibility of 'legitimate' input into wider social decision making. Even in non-democratic societies that do not offer such 'legitimate' channels, all citizens still confront fundamental questions of conformity or challenge, rebellion or revolt.

The differentiation of professional ethics from social ethics has also contributed to a situation where a lot of teaching and theorising about professional ethics is really teaching and theorising about rules and regulations of professional conduct set out in relevant statutes or in the codes of particular professional associations (of doctors, nurses, lawyers, researchers, teachers, managers etc.) or about the ways in which the general law applies to particular professionals in their relationships with clients. In other words, it is teaching about law rather than ethics (Parker and Evans, 2007).

More genuinely ethical consideration of particular professional practice needs to see such practice in a broader social perspective. Professionals within the discipline, and citizens outside it, should apply general principles of deontology and utilitarianism, of justice and fairness, in assessing the practices and principles of the discipline, and in considering, and campaigning for, possible reform.

Issues

Morality is difficult because of conflicts within and between these different sorts of values and commitments, as well as between individual self-interest and concern for others. The classic utilitarian hijacked aircraft scenario demonstrates such conflict between individual rights and requirements of social welfare. Similarly, our personal commitments tell us to put our own family and friends first. However, considerations of equal respect for the rights of all people and of support for the social good appear to conflict with this.

Individuals might believe that everyone should have a right of equal access to high quality education. But if parents find that private schools with high fees produce better results than public ones, if they can afford it, they would seem to owe it to their children to give them the best education money can buy. However, through paying such high fees, they are keeping such schools in operation and maintaining inequality and unfairness that denies such high quality education to children of poorer families. This would seem to be an area where government has the responsibility to create a situation where individuals don't have to make such choices, by ensuring that everyone has access to the highest quality education.

This example shows the need for concerted citizen action at the political level to allow for ethical decision making at the 'personal' level. It is far from being an isolated example. Such a need is absolutely pervasive in late capitalist society where working people have very effectively been turned against each other, competing for jobs and forced into complicity in the destruction of social welfare provision by prevailing social arrangements.

Another area of difficulty is the potentially complex relationship between ethical and factual or empirical issues touched upon earlier. An oft-repeated bioethical example is that of abortion. Killing people is wrong. But is a first trimester foetus a 'person'? This is a scientific question, rather than an ethical one.

Premise 1: killing innocent people is always wrong.
Premise 2: an early term foetus is an innocent person.
Conclusion: so abortion is always wrong, even in the first trimester.

Premise 1 is a clearly moral principle. A focus on human rights would put every 'innocent' person's right to freedom from being killed by others at the forefront of moral consideration. Article 3 of the UDHR states that 'everyone has a right to life, liberty and security of person'. Utilitarians, on the other hand, typically reject such a principle, since they believe that it can be necessary to sacrifice the lives of an innocent minority to save a majority.

But Premise 2 is a factual or empirical claim. However we might distinguish 'people' from other sorts of things as beings with conscious awareness, including feeling pain, a sense of self, or a capacity for rational thought, then it is a scientific question whether or not an early term foetus is such a being. If the facts prove that it is not, then abortion may be justified (under certain conditions) even while killing innocent people is always wrong.

What appear at first to be ethical questions often turn out to be questions about empirical facts as suggested by earlier considerations of utilitarianism. We need to be clear about relevant empirical facts in order to make good moral judgments.

Belief systems

According to historical materialist theory, human societies are organised in terms of three major sorts of emergent properties, functions or dimensions: the material or economic function, the ideological function and the political function. A human society exists only as long as particular relations, institutions and practices continue to fulfil these three functions. Social stability depends upon effective integration and mutual support of the three sorts of practices.

The material function is concerned with the reproduction of human material life through the production and distribution of material goods. The organisation and operation of the technology of material production and distribution through particular social relations of ownership and control of such productive resources constitutes the 'economic base' of society.

The ideological function is concerned with the mental reproduction of human life through the production and distribution of ideas, and the political function with overall social integration through the operation of law, policy and the exercise of coercive force. Together, ideology and politics constitute the social 'superstructure'.

Ideologies, in this sense, are organised systems of belief through which people come to identify themselves and make sense of the world they inhabit, through which they perceive and give meaning to the world. In contemporary society individual experience and action is mediated by a number of different ideologies, or belief systems, dealing with different areas of life, including beliefs about history, politics, the natural world, psychology and economics. Such ideologies interact, intersect, overlap and interfere with one another.

Ideologies have their own particular internal dynamics, principles of stability and change, growth and development. Particular disciplines of science, for example, advance through the application of particular sorts of research techniques, building upon a foundation of established theory. They are also influenced by external factors such as changes in other ideologies, and in the economic and political situation.

Such belief systems have a factual side and a value side. The factual side concerns ideas of what exists in the world today and how it has come to be, what has happened in the past and will or could happen in the future; what it is possible for human individuals and human groups to achieve. Clearly, various scientific (including social scientific) disciplines – and the popular understanding of scientific ideas – deal with these sorts of questions (of the existence of forces, atoms, black holes etc.). But political, philosophical, religious and legal ideologies also have an ontological dimension of factual claims, or assumptions, about the existence and nature of democracy, mind, god, crime etc.

All ideologies must make some reference to some real events or situations in order to become effectively anchored in social reality, and capable of directing perception and action. This does not preclude radical misrepresentation of reality or claims that imaginary things are actually real or real things imaginary.

The methods of science, properly applied, have been effective in discovering the true reality of nature and society beyond what is revealed to immediate experience. But even here, long established theories have ultimately been refuted by relevant observations. In areas where such methods are not applied so rigorously, or not applied at all, there is substantial scope for error.

The value side of an ideology concerns issues of inherent worth, of goodness, virtue and obligation; what sorts of things are intrinsically worth striving for (truth, happiness, freedom etc.); what sorts of characteristics make a good person (hard work, temperance, kindness, bravery, honesty etc.); what sorts of duties or commitments should be taken most seriously (to protect one's family, or one's nation, or one's class, to respect the property of others, to protect the shareholders); what ought to be, rather than what is the case. This includes ideas of what sorts of things are bad or wrong (torture, cowardice and communism) and what should never be done (slavery, murder, incest).

If the value side of a belief system accords high priority to protecting or attaining a particular sort of thing (e.g. freedom, communism or God's love), then we would expect the factual side to provide what looks like evidence for the existence or possible existence of that kind of thing (that it actually does or could exist) as well as guidance as to how to protect or attain the sort of thing in question. Typically, fact and value issues are deeply intertwined within particular religious, political and economic belief systems.

Individuals acquire their moral ideas of right and wrong as elements of particular ideological constructions, in the course of growing up, from their families, friends, neighbourhoods, from the media, from social groups, from workplaces and religious and political organisations, and from formal and informal education. There is reason to see such beliefs as grounded in or built upon innate tendencies, drives and potentials inherited from our evolutionary ancestors. However uncritically such beliefs might originally have been absorbed or embraced, they subsequently become reinterpreted, transformed or replaced as a result of critical thought processes and life experiences.

Class inequalities

Corresponding to the three dimensions of social organisation are three main sources of power and domination in human society: control over the major means of material production, land, raw materials, technology and labour; control over the major means of communication, education and persuasion; and control of the means of political administration and coercion (Miliband, 1991, p. 27). Where such sources of power are unevenly distributed, where particular elite groups utilise concentrated material, ideological and political power to dominate and control the lives of others, materialism identifies the society as class divided.

Historical materialism highlights the central role of people's material life circumstances, and the interests associated with their particular positions within a social division of labour, in determining which ideologies make sense to them, how they respond to, interpret, develop and transform existing ideologies and create new ones.

In a class-divided society, those enjoying power and privilege denied to others will tend to endorse and embrace ideas which support such power and privilege. They will be in a strong position to ensure the further development and wide propagation of such ideas. This includes ideas which can practically assist them in maintaining and extending their power, as well as ideas which serve to justify and legitimate such extension through presenting it as natural, inevitable, well-deserved or optimal for all concerned. Ideas which help them to effectively resolve their own differences without undermining their solidarity as a class will contribute to the effective perpetuation of such power and privilege.

As noted above, there are reasons for believing that ethical experiences and intuitions are grounded upon biological realities; that feelings of empathy and intuitions of justice and fairness might be structural features of a biological human nature, derived from animal ancestors, rather than purely contingent historical realities. This could go some way towards explaining why those enjoying power and privilege, at the expense of others denied these things, need ideas which also serve to assuage guilt and rationalise oppression and inequality.

For those suffering ongoing oppression and disempowerment, ideas which challenge status quo are likely to make more sense, such as ideas which can help to explain the weaknesses of existing social structures and relations, which can provide practical assistance in reducing inequality and oppression, or in dismantling existing relations of domination and replacing them with new and better structures and relations.

Elite groups' control of surplus material wealth, over and above the requirements of material reproduction of the social productive forces, facilitates their effective control of both the exercise of coercive political power and of the production and distribution of ideas through control of major means of communication, education, research etc. Through such control, they can shape the ideas of the disempowered to counteract or preclude the development of effective counter ideologies. In particular, the disempowered can be persuaded to accept such elite groups' own ideas of the benefits or the necessity or the inevitability of such inequality. They can be systematically distracted from serious consideration of their situation, or the true facts of that situation can be systematically misrepresented and mystified. Typically, all of these elements are involved.

Control of major communications channels by particular elites does not imply absolute power to prevent the survival and development of counter ideologies of the oppressed. The more powerful such counter ideologies become, particularly in times of economic crisis, when the power of such elites is threatened, the more the dominant ideologies have to acknowledge and address such alternative views. Different elite power groupings, fighting for a dominant position, can develop their own variations on the dominant ideology to direct their struggles and win the support of other class groups. Thus, ideology, along with the economic base itself, becomes an arena of ongoing class struggle.

Liberalism

In a capitalist society, as in any other class-divided society, the dominant ideology, other than in times of serious breakdown of the social order, is an ideology produced, propagated and endorsed by those exercising economic, political and ideological power and by those serving them. It functions to legitimate and support dominant capitalist relations of production. As elsewhere, different currents within such an ideology represent different currents of thought and different priorities within the powerful elite groupings and different degrees of engagement with, or concessions to, counter ideologies, challenging the existence of capitalist class inequality.

The dominant social ideology of the English-speaking world, which shapes beliefs and perceptions relating to all areas of social life, is liberalism. This is not to be confused with the beliefs of the various contemporary political parties that call themselves 'Liberals' or with the left side of mainstream politics in the US, though both are influenced by liberal thinking. Rather, as Peter Self (2000, p. 34) explains, 'liberal thought', in this context, 'refers to the pursuit of a coherent body of thought about political, economic and social relationships, founded on some basic concept of the worth and entitlements of each individual'.

While there are many forms and shades of liberal ideology, all liberals 'celebrate the rights and responsibilities of individuals'. Liberalism's

> strong attachment to individual freedom stresses the right of individuals to choose and make their own lives and to take responsibility for the outcomes. It also endorses equality in the sense of the equal moral worth of all individuals and their entitlement to respect and consideration ... [it includes] a strong belief in the value of human reason and in the possibility of human progress. (Self, 2000, pp. 34–5)

Liberalism arose in opposition to

> monarchical power, feudal privileges in land and offices and arbitrary government ... In their place was put constitutional government, meaning the rule of law, an independent judiciary, representative assemblies and civil rights ... [where the latter include] freedom of opinion, freedom of assembly for peaceful purposes, freedom of worship, freedom from arbitrary imprisonment. (Self, 2000, pp. 34–5)

Most liberals believe in some forms or degrees of representative democracy, democratic rights and the rule of law. They support an effective separation of executive, legislative and judicial powers. But different groups have different ideas about what democracy and law could or should be, about the nature of people and society, justice and fairness.

Most liberals see a central role for market relations in a free and efficient society. They support the 'individual right to own, inherit and dispose of property, to choose one's own occupation and to profit from one's own abilities' (Self, 2000, pp. 34–5). But separate strands of liberal thinking are distinguished by different ideas about capitalist market relations and market forces. There are varying ideas about the extent to which such market relations and private property in the means of production need to be regulated and restricted by state power, and complimented by public ownership and planning.

Disparate strands of liberal thinking involve different forms and degrees of deontological and utilitarian ethics, contrasting ideas of the meaning and significance of human rights and social welfare, of freedom, justice, and equality. In all cases, such ideas are intimately intertwined with particular factual beliefs, and assumptions about people, society and history. In particular, liberal ethical ideas are bound up with economic theories and ideas about the nature of capitalist market relations.

Right-wing liberalism

At one end of the scale are right-wing or free market liberals, seeking to minimise any kind of public regulation or restriction of private property and free markets, including individual or corporate ownership of major productive forces. Some right-wingers defend these institutions in utilitarian terms, as the most efficient possible forms of organisation of production and distribution, generating the highest possible levels of social wealth and hence of (at least potential) social welfare. In particular, private ownership and free markets are seen as intrinsically much more efficient wealth generators than state control and planning. Genuine concern for social welfare, therefore, requires that such state intervention be kept to a minimum.

Right-wing or free market liberals believe that people are basically selfish but also generally rational. People are competitive and power seeking and continually calculate the costs and benefits to themselves of the different actions available to them. They choose actions that maximise immediate pleasure or minimise immediate pain, or that promise to maximise pleasure in the future. Costs include such things as hard work, sacrifice of time or effort, risks to person or property, and discomfort of one sort or another. Benefits include money as the means to acquire desirable goods and services, power over others, comfort, pleasure, recognition and protection from threats to general well-being.

Right-wing liberals see capitalist market relations, with appropriate criminal and civil legal support, as the ideal and indeed the only mechanism capable of effectively harnessing individual greed and competitiveness to serve the common good. In a free market, productive innovation, efficiency, training for special skills, risk, effort and quality of output are rewarded and encouraged with higher revenues, profits or wages, while prices to consumers are kept as low as possible by ongoing competition between producers. People thereby help others by helping themselves. Failure to contribute, or time and effort wasted in unproductive endeavours, by contrast, are rewarded with poverty and privation. Such privation also has its utilitarian place, as an incentive for all to work harder and better so as to avoid it.

Many right-wingers argue that free markets not only generate the greatest possible social wealth, they also generally ensure a fair distribution of such wealth. Markets are seen as fair, as they generally ensure appropriate rewards to labour, capital and entrepreneurial initiative in proportion to the actual wealth created by such factors of production. The better someone is in creating social wealth, through the exercise of special abilities, through showing initiative and taking risks, or introducing significant innovations, the more they sacrifice in terms of painful education and investment, or long hours of hard work, the more they can expect in terms of ultimate reward in profits, revenues, wages, salaries, bonuses or royalties.

Right-wing liberals recognise 'human' rights to life, to ownership of legitimately acquired private property, and to take steps to protect such life and property. They recognise democratic rights to political organisation and participation. They see legitimate nation states as essentially products of the action (and interaction) of free individuals, contracting together to create an effective means of protection of their lives and properties, through the exercise of coercive central power.

Beyond this point, right-wing liberals see no necessary requirement for individuals or for central authority to take steps to protect or enforce any other rights per se. The marketplace is seen as the arena for the free negotiation of other such rights and responsibilities, where individuals contract together for the supply of particular goods or services for what are seen as appropriate payments. If people want to exercise their 'rights' to education or health care, for example, they must expect to pay for others services in these areas. Only if such providers renege upon such freely negotiated contracts will the state intervene to enforce the contractual rights in question.

Whereas left-wingers see a government responsibility to take action to ensure universal employment and education, fair wages and health care, right-wingers see such rights in terms of individuals' freedom to pursue their own arrangements through the operation of free market forces. No one has a right to prevent people buying and selling educational and health services, from offering or taking on whatever employment they choose with whatever wages they choose to offer or to accept. Equally, no one has any responsibility to provide any such things to other people unless they take on such a responsibility through formal contract or informal promise (e.g. to family members).

Nor can anyone expect that their life or property right, or rights of political participation, will necessarily be protected or respected if they fail to respect the life and property of others. Rather, state power itself reserves the right to take away life, property and democratic rights from those deemed unworthy by virtue of serious legal transgressions.

Right-wing liberals argue for social arrangements which maximise the scope of free human choice and action, to allow individuals to find their own fulfilment in their own way. In particular, individuals should be free from restriction and control by coercive state power. Optimum efficiency and/or fairness in market operations are seen to depend upon individuals being as free as possible from political control. Beyond a basic respect for others' life and property, people should be morally bound only by their own freely chosen commitments to others. This will typically include commitments to close family members – whose interests are thereby equated with individual interests.

Coercive legal sanctions should only be applied where individuals directly infringe upon the life, bodily integrity, legitimately acquired property or free action of others. Generally, liberal right-wingers see the need for serious punitive intervention in such cases, in the interests both of retribution and of specific and general deterrence.

This is the core role of the state – to exercise an effective monopoly control of coercive power in order to protect the life and property of citizens from internal and external threat. The criminal law, including the use of police, courts and prisons, protects citizens from fellow citizens, while the armed forces protect them from military threat from abroad. Representative democracy – with periodic changes in the political leadership in response to popular decision – prevents the use of such a monopoly of coercive force extending beyond such protection of life and property to become a source of totalitarian tyranny.

State power receives its legitimacy through its effective protection of basic human rights – to life and property – and through the citizens' ongoing consent to its authority, demonstrated through their continued acceptance of the benefits it bestows.

While effective protection of life and property has to be paid for through some system of taxation of legitimate private wealth, right-wingers are hostile to any further taxation to support the sort of comprehensive state welfare provision favoured by more left-wing liberals.

Liberty versus justice

Major right-wing liberal philosophers have been hostile to any principles of justice they have seen as conflicting with rights to enjoy, employ and dispose of legitimately acquired private property. They have identified justice with the operation or enforcement of freely chosen contractual arrangements.

Classical liberal philosopher John Locke argued that individuals have a natural right to appropriate unowned goods by transformation of natural resources into humanly useful products through labour, provided that this does not significantly diminish others' opportunities to engage in such appropriation. Locke saw this principle as allowing 'original' appropriation by some to deprive others of the possibility of such appropriation providing that it opened up other, equal or better opportunities for those others (Levine, 2002, Chapter 3). Appropriation of limited land area within a territory, for example, can be justified if it provides good enough employment opportunities for those no longer able to achieve such original appropriation.

Following such appropriation, individuals have a right to do whatever they want with such legitimately acquired goods, provided they do not infringe upon such legitimate property rights, and rights to life, of others. As well as enjoying and employing such goods in whatever ways they choose, they can rightfully engage in market exchange and gift giving, including bequests, insofar as such exchanges are free and uncoerced and involve no fraudulent misrepresentation. Those who acquire goods through such exchange thereby become the legitimate owners of such goods, whatever particular patterns of ownership might be produced by such processes.

Twentieth century libertarian philosopher Robert Nozick built upon a Lockean foundation to criticise all ideas of distributive justice based upon any sort of overarching pattern or principle, including ideas of justice as desert or equal apportionment. Rather than protecting others' opportunities for wealth acquisition, he argues that initial appropriation of natural resources is justified if others are left as 'well-off' as they were before. So, for example, miners can legitimately benefit from mining operations in particular territories providing they leave others in the vicinity no worse off than they were originally.

As there are not many unowned objects, materials or spaces left in the modern world, legitimate acquisition through exchange or gift is crucially important in this theory. Whatever pattern of property ownership emerges from such operations is legitimate or just insofar as it is the product of free and legitimate transactions, including wage contracts and inheritance. State powers have no rights, or very limited rights, to 'interfere' in such processes through coercive appropriation in the service of welfare, redistribution, equality (of opportunity or outcome) or just deserts.

Nozick argues that allowing individuals the ability to engage in such free exchange will inevitably undermine any 'distributive pattern principle', including any principle of 'just deserts'. Any such principle could only be enforced through continuous monitoring and intervention in everyone's lives by a totalitarian state power, which is fundamentally unacceptable. Private ownership is an inalienable right and rights claims and legitimate contracts always take precedence over other considerations. Where concrete 'liberty' to exchange, bequeath or inherit comes into conflict with abstract ideas of justice or equality, liberty should always win.

Law and justice

Despite Nozick's hostility to patterned principles of justice, as noted earlier, many right-wing liberals still do see justice, both in distribution, and particularly in retribution for criminal acts, in terms of just deserts. They see it as ethically legitimate that individuals should be rewarded in proportion to the monetary value of their positive social contribution as measured by the price people are prepared to pay for the goods in question. Individuals should be subjected to painful sanctions in proportion to the extent of the harm they cause – not just harm to individuals but also harm to the rule of law itself – and to the capitalist market system.

The capitalist market system is seen to (generally) deliver such distributive justice, while the common criminal law (generally) delivers retributive justice, through the application of appropriate punishments for serious wrongdoing.

Civil law, meanwhile, allows free individuals themselves to take action against those who have (intentionally or negligently) trespassed upon their person or their property or failed to honour contractual commitments. The main principle here is that of restitution – restoring the wronged individual to the state they would have been in if they had not suffered the harm in question – to person or to property.

The market system inevitably produces inequality of outcome. Some end up richer, more powerful and more respected than others. Many right-wingers see this as a consequence of just deserts. As noted earlier, some also seek to justify such inequality in utilitarian terms, as contributing to overall social welfare. At one end of the scale, substantial rewards ensure that the most talented people use their talents to the best advantage of all through innovation, investment, entrepreneurship, and the exercise of special – professional – skills. At the other end, the miseries of poverty and deprivation are necessary to overcome the inherent laziness of some – or all – humans. Economic inequalities fuel aspirations for more consumption goods which fuel economic growth.

Due to the centrality of the market and the market value of goods and services in right-wing liberal thinking, growth in the gross domestic product (GDP) (in the total value of goods and services produced in a given year within a country) per capita compared to the previous year, is taken as an important objective measure of increasing social welfare. The higher the GDP per capita, and the greater the rate of its growth, the greater the overall social well-being.

Here we see the need for much more detailed consideration of the actual nature and operation of capitalist private property and market forces – as portrayed in liberal ideology and as existing in reality – in order to properly understand and evaluate the right-wing liberal position. Such detailed consideration is the subject matter of the next two chapters.

Left-wing liberalism

It was earlier noted that there will typically be different currents of thought within broadly dominant ideologies, representing the different perspectives and interests of different elite power groupings. Different currents will also reflect different responses to changing social circumstances and developing counter ideologies. Left-wing or 'social' liberalism arose and developed as a response to deepening crisis in the capitalist market system and to the development of socialist ideas as a major challenge to capitalist production relations.

Left-wing or social liberals recognise a deeply social side to human nature which motivates people to see their own identities and interests as intrinsically bound up with those of others. Whether people are more or less selfish and competitive depends upon the social situations in which they grow, develop and live. This includes involvement in institutions and ideas which encourage and endorse or discourage and condemn such selfishness and competition. Social liberals aim to create social relations that encourage cooperation rather than competition, in part through reducing the scope of market relations.

Social liberals have a less optimistic view of the efficiency and fairness of market relations. They argue that success in the market can have as much to do with luck and ruthlessness, exploitation of monopoly power, cost externalisation and criminality, as with reward for genuinely valuable social contribution, effort or sacrifice. Left to themselves, free markets produce increasing – and undeserved – inequality of wealth and power, monopoly and unemployment, slumps and depressions, pollution and ill health, which undermine democratic institutions and practices and radically fail to maximise social welfare.

Undeserved inequality and poverty restrict the freedom of action of the relatively powerless and poor. It is their social circumstances of restricted opportunity, rather than their free choices, and cost–benefit calculations, which explain their actions, including resort to crime in some cases. It is the social conditions that need to be changed to reduce such inequality and such crime.

Social liberals support equality of opportunity, but recognise the practical difficulties of achieving it and the need to compensate for failure to achieve it through various 'countervailing measures' of redistributive taxation, welfare and positive discrimination in some cases.

Social liberals believe in rewarding genuine effort and sacrifice in valuable social contribution. For social liberals, justice also requires public action to improve the lives of the less well-off, and sympathetic treatment of their failure to make a positive contribution. Those benefiting significantly from such inequality should be held to a higher standard of accountability for law breaking, and should be legally bound to contribute towards reduction of poverty and inequality.

In contrast to socialists, who call for the complete dismantling of market relations in favour of a fully planned world economy, social liberals believe that the market can be 'tamed' and humanised through appropriate and extensive government intervention and regulation. This will typically require direct government control of key productive and financial operations, with direct investment to maintain full employment, centralised collective determination of workers' wages and conditions with substantial trade union input, heavily redistributive taxation to fund comprehensive social welfare provision, including high quality public health, education, housing and social security.

They emphasise the responsibility of state power to take an active role in ensuring a basic equality of satisfaction of the material and social needs of all. Whereas right liberals defend civil and political rights, left liberals also focus upon the responsibility of public authority to ensure fulfilment of economic, social and cultural rights as set out in the UDHR. They believe in helping or rehabilitating, rather than punishing, people whose social situation makes it more difficult for them to contribute, or predisposes them to criminal acts.

Rather than GDP, social liberals look to quantitative data on reductions in inequality, ill health and crime, and increasing levels of education and democratic participation, as objective measures of social well-being and progress. Here again, as with the right liberal position, we see the need for a more in-depth consideration of economics in order to properly understand and evaluate the left liberal position, including the critique of the right wing. Again, such consideration is provided over the next two chapters – particularly Chapter 3.

Rawls' theory of justice

Just as politicians, teachers, commentators and others of right- and left-wing liberal commitment call upon both teleological and deontological ethical ideas to clarify and justify their political views, policies and programs, decisions and actions, so do political philosophers conjoin both right- and left-wing liberal ideas in the development of particular ethical theories. This is clearly illustrated in John Rawls' massively influential 'contractarian' theory of distributive justice, seen by many as 'the most important contribution to political philosophy in the twentieth century' (Norman, 1998, p. 190). Anyone investigating the relations between ethics, liberal politics and justice will inevitably encounter reference to the key ideas set out in Rawls book *A Theory of Justice*.

In this book, Rawls addresses the key question of 'how rights and duties, benefits and burdens' can be 'fairly distributed between members of society' (Norman, 1998, p. 190). He rejects the idea that utilitarianism has anything to contribute here, since its aggregation of the happiness or well-being of all the individuals in the society concerned completely fails to take account of fair treatment of every such individual.

To cast light upon this basic issue of distributive justice, Rawls appeals to what he calls an 'original position' in which individuals are imagined to be deciding upon principles of justice to govern the operation of a future society in which they will live. Such a decision is made behind a 'veil of ignorance', in the sense that such individuals do not know what their position will be in such a future society. They don't know what particular talents and abilities, or disabilities, they will have (Norman, 1998, p. 190).

In key respects this is a return to classical right-wing liberal contract theory, as considered earlier. In contrast to later, social liberal recognition of the intrinsically social and relational character of human identity, seeing individuals as integrally embedded in relations of 'family and friendship, of work and culture and political life, and in ties and loyalties which transcend the individual self' (Norman, 1998, p. 193), Rawls appeals to a 'pre-moral, pre-social individual', the rationally self-interested individual of classical liberal theory.

His aim is to show how a practical concern for others can be generated out of such original selfishness as individuals are forced to consider the interests of others through the device of the veil of ignorance. They are forced to consider the position of others suffering serious social disadvantage – as a consequence of lack of inherited wealth or intelligence or ambition, racial or gender discrimination etc. – through considering the possibility that they themselves could be such people.

Rawls argues that in consequence of this, individuals in the 'original position' would choose to organise their future society on the basis of two key principles:

1. Each person is to have an equal right to the most extensive basic liberty compatible with a similar liberty to others.
2. Social and economic inequalities are to be arranged so that they are both (a) to the greatest benefit of the least advantaged and (b) attached to office and positions open to all under conditions of fair equality of opportunity (Rawls, 1972, p. 60, p. 83).

As Norman (1998, p. 190) explains, Rawls expands and illustrates the first principle by reference to a typically liberal emphasis upon protection of democratic rights to 'freedom of thought and conscience, freedom of speech and assembly and the right to vote and stand for office'.

The second principle focuses upon the utilitarian idea of individual well-being, but requires that social arrangements be adopted that make everyone as well-off as possible, rather than maximising the sum total of such individual well-being.

> The test of this is whether inequalities are to the advantage of the least well-off. Arrangements which make some people better off than others can be justified only if they make even the most disadvantaged better off than they would otherwise have been. Only then can it be said that this is a state of affairs which everyone could agree to, and which does not sacrifice anyone's interests for the benefit of others. (Norman, 1998, p. 191)

Norman is critical of Rawls starting with 'pre-moral and pre-social' individuals, maintaining that 'we cannot understand the nature of moral values and why they matter to us' unless we recognise that human identity integrally involves social and moral relations (Head and Mann, 2009, Chapter 9). He argues that even if it were true that people in the original position would agree to such principles, Rawls provides no good reason why real people in the real world should accept such principles.

Principle 2, which Rawls calls 'the difference principle', is not necessarily associated with a right-wing liberal perspective. However, Rawls himself develops it by reference to the idea of financial and other incentives supposedly necessary to ensure that work involving special training and responsibility is appropriately carried out by those best able to perform it. While, in fact, the difference principle is supposed to be empirically based, it is all too easy for right-wingers to appeal to it in support of an existing status quo of radical inequality without any serious consideration of possible alternatives.

But while Rawls does indeed argue strongly for 'justice as contract', so does he develop another idea, of 'justice as co-operation', much more in tune with the left liberal emphasis upon participatory democracy (Norman, 1987, Chapter 4). As Rawls says

> The intuitive idea is that since everyone's well-being depends upon a scheme of co-operation without which no one could have a satisfactory life, the division of advantages should be such as to call forth the willing co-operation of everyone taking part in it. (1972, p. 15)

This can be taken as recognition of the fact that it is just as important to consider the actual process of formulating and applying principles of distribution in real social situations as to consider the nature of such principles in deciding whether they are, indeed, fair and just. An authoritarian central state power could require and enforce some form of just deserts, for example, but there remains the question of whether any such system could really be just if it were imposed from above.

As Norman (1987, p. 74, p. 78) argues, cooperation 'is a form of association in which people work together voluntarily, respecting one another as free participants in a common enterprise'. It can be distinguished from coercion where some are forced to follow others' orders and from exploitation, where one group uses its coercive power over another to further its own interests at the expense of the other. So it can be distinguished from competition, where some strive to do better than others and thereby benefit at others' expense. Genuine cooperation therefore presupposes an equality of power within the whole group, to preclude any such differential benefit at another's expense.

Norman argues that such genuine cooperation in establishing principles of distributive justice will probably ensure a fundamental equality in such principles, in the sense that

> if all members of the group are committed to working together co-operatively, they will see to it that everyone's interests are taken into account and no-one's interests are sacrificed to those of others... The set of arrangements which is adopted... must be one which can be – rationally – justified to each of its members. If a co-operative group is one in which each person has an equal say, then each of them can make an equal claim on the group, those claims can be properly satisfied only by a state of affairs in which all benefit equally overall. (1987, p. 70)

Many right-wingers argue that the closest we can practically get to any such equality of power in a significant decision situation is a system of representative democracy in which everyone's vote counts equally. Some left-wingers focus upon how far removed current democratic voting processes are from any genuine cooperative, ongoing, participatory deliberation and collective formulation and application of principles of distributive justice. They seek to reform social institutions and practices in ways which extend the scope of such genuine cooperation, with more people involved in more effectively cooperative decision making shaping more and more aspects of social life, and particularly issues of fair distribution of social benefits and burdens.

Later chapters

Mainstream politics in the Western world has oscillated between more right and more left liberal ideas and policies. We are just emerging from 30 years of right-wing – neoliberal – domination. In part, such domination has rested upon – and has further contributed to – widespread acceptance of key elements of right-wing ideology and right-wing ethics. In this context, it is particularly important to see precisely how these ideas work, how they interface with public policy and with reality.

Later chapters consist, essentially, of an exploration of the consequences of the application of neoliberal ideas in areas most directly affecting human survival, health and well-being. They begin to develop a social liberal critique of such application and the beginning of some proposals for reform.

DISCUSSION TOPICS

1. What are human rights? Is there a human right to good health?
2. What are the major advantages and limitations of utilitarianism?
3. What is justice?
4. How do right- and left-wing liberalism interpret and apply ideas of human rights, utility and justice?

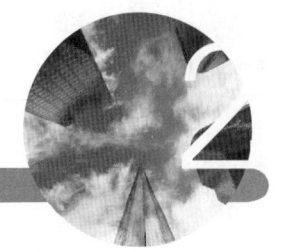

Freely competitive markets

As noted in the last chapter, liberal ideology and liberal ethics, including liberal approaches to issues of biomedical ethics, cannot be understood or evaluated without understanding some of the key economic ideas and developments which have shaped and underpinned liberal thinking. This chapter looks at the economic underpinnings of contemporary right-wing liberal ideology. It provides an overview of key ideas behind neoliberalism and its policy prescriptions.

A key element of such neoliberal economic ideology is the idea that the interests of economic welfare are best served by privately owned corporations driven by the goal of profit maximisation, competing with each other for customers in a free market; that capitalist markets serve as a means for harnessing universal human self-interest to serve the common good through competition. These ideas are supported by a number of influential economic theories developed over the last 200 years.

Smith, Marshall and Hayek

Adam Smith

The classical defence of free market competition goes back to Adam Smith's (1723–90) writing in the late eighteenth century. According to this analysis, competition benefits consumers by forcing enterprises to keep the selling price of their goods down and the quality of such goods up in pursuit of increased business and profit. Partly through the operation of a capitalist financial system, the profits generated can then be reinvested in the production process, continuously expanding productivity, output, income and wealth.

In the *Wealth of Nations* (1776), Smith argued that increasing mechanisation and division of labour were the foundations of economic growth. He saw how increasing division of labour for the development and enhancement of specialised skills saved time in moving between tasks and facilitated the development of labour-saving devices allowing more to be produced with less time and effort (Foley, 2006, Chapter 1; Pressman, 1999, p. 21; Rubin, 1979, Part Three; Stilwell, 2002, Chapters 8, 9 and 10).

Smith argued that the sale price of a good is determined by its cost of production, with its 'natural price' 'being the sum of costs of paying land, labour and capital for their role in production' (Pressman, 1999, p. 24). This is the equilibrium price, towards which market prices move. If increased demand pushes market price above this, thereby increasing profit (as revenue above the equilibrium price) in a particular industry, then producers shift their resources into that industry. Increased supply and competition amongst such producers then reduces market price bringing it back to the natural price. Similarly, if market prices fall below the natural price, due to reduced demand in a particular industry, producers shift their efforts elsewhere, leading to reduced supply, increased competition amongst consumers and a rise in market prices back up to the natural price (Foley, 2006, pp. 14–15).

Each of the factors of production was to be paid at its 'natural' rate. At times, Smith focused on rent as a monopoly price for the scarcity and non-reproducibility of land. He saw profit as a reward for saving money to provide capital for investment, without explaining its natural rate. He saw the natural wage as the minimum needed to provide for workers' subsistence. If wages fell below this, workers and their offspring would die and with employers competing for fewer workers, wages would rise. If wages rose above subsistence, more workers' children would survive and increased competition for jobs would drive wages down again (Foley, 2006, pp. 20–2).

In a competitive market, those producers who can improve their productivity through increased mechanisation and division of labour can produce more goods for less. They can thus make more profit and increase their market share, selling below the natural price. This forces other producers to adopt similar innovations in order to stay in business, thereby reducing the natural price and bringing market price back into line with the new natural price through competition amongst the producers.

This is how the 'invisible hand' of the market benefits all consumers; without any kind of concerted planning process, the self-interest of producers drives them to respond to changing conditions of demand, shifting their efforts to where they are most wanted, and all the time improving their productivity, cutting costs and delivering cheaper goods to consumers. Different regions and different nations will have different natural prices for particular goods by virtue of different conditions of nature and society. Free international trade forces regions and nations to concentrate their efforts in areas where they have a competitive advantage by virtue of lower natural price, and gives all nations as consumers the benefits of lowest possible market prices (Pressman, 1999, p. 22).

Clearly, this system only works so long as there are no barriers to the free flow of resources from less to more profitable employment. If producers in a particular – more profitable – sector can stop others entering that sector, then they can stop competitive price reduction through increased supply in that sector, and thereby also avoid the need for further innovation to bring their costs down. Monopolies thus lead to higher prices for consumers, reduced pressure for good management and innovation in the monopoly sector, and 'misallocation of resources' (Pressman, 1999, p. 23). The higher profits of the monopoly sector also increase the political power of the producers in that sector to pressure government to serve their particular selfish interests by protecting and extending their monopoly position. This includes the undermining of free international trade by tariff protection of the sector in question.

Alfred Marshall

A somewhat different sort of defence of free market competition was provided by Alfred Marshall (1842–1924), the founder of neoclassical partial equilibrium analysis of individual markets which now forms the core of microeconomic theory. In contrast to the supply driven approach of Smith and the other classical economists, Marshall emphasised the interaction of supply and demand in the determination of prices and production (Stilwell, 2002, pp. 148–50).

He argued that 'consumers are forever attempting to get the greatest utility from what they purchase and consume' (Pressman, 1999, p. 65). When a good is highly priced they will tend to buy little of the good because of the greater utility from using the money to buy many other things (Pressman, 1999, p. 65). This is the foundation of the law of demand – as prices rise, consumers buy lesser amounts of a good, leading to a downward sloping demand curve – with price on the y axis plotted against quantity on the x axis.

Marshall agreed with Smith that supply is determined by production costs. In this case, he argues that beyond a certain point of returns to scale, output can only be increased with rising costs (per unit of product). So producers' profits can only be maintained if they are able to charge a higher price for each unit of such increased output. 'Hence the Marshallian supply curve is positively sloped [rising from left to right]' (Pressman, 1999, p. 66). 'The two scissors of supply and demand determine the price for each good and the amount of each good that will be produced' (Pressman, 1999, pp. 64–5).

Marshall argued that 'competition forces actual prices towards the equilibrium price' (Pressman, 1999, p. 65) – where the supply and demand curves intersect. Higher than equilibrium prices lead to unsold goods, telling firms to lower prices and cut production. Lower than equilibrium prices lead to shortages, telling firms to increase prices and raise production. Only at equilibrium do firms sell all they produce at stable prices. No goods are wasted through failure to find buyers and no purchasers are denied access to goods at prices they are willing to pay.

Marshall understood demand–change as change in the amount of goods purchased at a particular fixed price. Such shifts in the demand curve, forwards or backwards, result from changes in income, wealth, taste or population, from changes in the prices of other goods or from changed expectations of future prices. Changes in the number of goods supplied at the same price result from changed production costs. With higher costs (e.g. produced by higher wages) businesses can make the same acceptable level of profit only by passing on such higher costs in the form of higher prices. New technology can lower unit costs insofar as it means less labour is necessary to produce each unit of output, allowing the same level of profit at lower sales prices (Pressman, 1999, p. 66). Competition will tend to produce such reduced costs as businesses seek surplus profits over and above such acceptable levels through reduced costs and increased market share. It will tend to produce reduced prices, as all producers are forced to adopt the new and improved technology.

In this scheme, wage levels are determined by the interaction of household supply, of different skill levels and types of labour, and business demand. Marshall believed that increased mechanisation continually reduced the demand for unskilled labour, 'keeping wages down for the unskilled' (Pressman, 1999, p. 67).

Friedrich Hayek

The work of Smith and Marshall suggested the need for government intervention to prevent the formation and operation of monopoly market power through bankruptcies, mergers and takeovers, or through price fixing agreements between different business operations. Marshall called for progressive taxation to help the poor. The work of Friedrich Hayek (1899–1992) between the 1940s and 1970s, by contrast, provided a more extreme defence of free market competition without the need for any such government intervention, which appealed particularly to the neoliberal political reformers of the 1970s and beyond.

Hayek attributed virtually all economic problems to government interventions of one sort or another. He argued that the complexity of a modern market economy was such that no government authorities could keep track of the workings of the whole economy – understood as a web of different supply and demand conditions (Pressman, 1999, p. 118). Such authorities could not, therefore, rationally plan and intervene in such market operations, with any attempt to do so actually undermining market efficiency (Foley, 2006, p. 153).

Rather than seeing a need for government intervention to preclude monopoly power, Hayek argued that monopolies are typically produced through government intervention, as where a domestic producer lobbies for restrictions on foreign imports.

> Hayek also thought that even if large firms become powerful, potential competition [or the threat of new rivals starting up] would force firms to operate efficiently and produce the goods demanded by their customers at the lowest possible cost. (Pressman, 1999, pp. 118–19)

Issues of justice do not apply to income distribution by 'impersonal' market forces. Attempts by government to redistribute income will inevitably make things worse for everyone. Taking from the rich to give to the poor reduces the incentives for anyone to generate further wealth.

> This leaves less for everyone, wealthy and poor alike. The poor are also hurt because the wealthy perform important economic functions like taking risks . . . and testing new and expensive products that, if successful, get mass produced at lower prices. (Pressman, 1999, p. 119)

Neoliberals typically draw upon elements of all three of these approaches to justify more or less 'laissez-faire' policies of free trade and investment. Whereas others advocate various kinds of selective industrial policy, favouring the development of certain industries over others with a view to enhancing economic welfare in the long run through a range of different interventions, neoliberals argue that the state should not seek to shape industrial development. Political considerations inevitably induce inefficiencies. The market mechanism alone is capable of optimally allocating economic resources, outside of a few exceptional areas.

Corporations – private ownership of productive resources

Neoclassical theory centres on the idea of 'perfect competition' where there are a large number of medium-sized business enterprises, with none big enough to gain any special advantages. Each business accounts for only a small proportion of total output for the industry, their outputs are homogenous and substitutable, and there is no restriction upon the entry of new producers. Businesses have no significant power to determine the prices of goods, they can only make decisions about the quantity of goods they will produce; these are preconditions for 'efficiency' in the neoclassical sense of the term. This concept of 'efficiency' is supported by ideas of the optimal firm as being relatively modest in size. In theory, diseconomies of scale and size typically set in before firms can acquire any significant degree of market power.

Despite this, neoliberals generally see large-scale private business corporations as appropriate organisations of control of major productive forces, with ownership of corporate assets by private shareholders. Support for public corporations goes along with support for portfolio investment as the most efficient form of funding of productive activity.

The professional and hierarchical structure of modern corporations allows for the mobilisation and integration of the work of experts, directed by a central authority, in effective pursuit of profit maximisation. Large scale allows for economies of scale and concentration of resources for large-scale projects, which sustain further scale economies. It allows for substantial research budgets, increased division of labour, specialisation and automation of production, along with bulk purchase, storage and application of inputs to bring down costs. It also facilitates overseas expansion to gain direct access to overseas resources, including cheap labour and raw materials, and 'supportive' political systems.

These considerations – potentially – apply to state owned and run enterprises also. But neoliberals argue that such gains can only be effectively realised through private ownership. In particular, they argue that because managers of state owned enterprises are not owners, they lack the sorts of incentives of increased profit and power that motivate owner–managers in the private sector to ensure productive efficiency (Chang and Grabel, 2005, p. 83).

It's true that management in big public corporations is separate from ownership. But

> management performance is monitored by shareholders and checked by market incentives and competition. Stock options, which give managers an ownership stake in the firms they run, also strengthen the incentives for them to perform well . . . the liquidity of capital markets gives shareholders the ability to penalise firms [and managers] for poor performance by selling their shareholdings. (Chang and Grabel, 2005, p. 83)

Managers can be sacked when firms are taken over by others acquiring their stock. The threat of such takeovers, along with pressures of competition for market share, keeps managers on their toes. By contrast, there are generally no shareholders to hold the managers of state owned enterprises (SOEs) to book, and many state owned enterprises are monopolies, protected by law. This contributes to SOEs producing shoddy products and providing low standards of service while maintaining high prices (Chang and Grabel, 2005, pp. 83–4).

We can now understand the superiority of share markets to banks in funding production as they can more directly reward or punish managers' performance, and they are free from restrictions imposed by the banks. Portfolio investment also 'promotes the diffusion of risk and thereby financial stability and investment through the wide dispersal of asset ownership on capital markets' (Chang and Grabel, 2005, p. 125). Limited liability and the capacity for a speedy exit, provide the protections to allow the public to invest widely and deeply to sustain economic growth and development.

Privatisation and intellectual property

The superiority of private ownership of productive resources leads neoliberals to promote the movement of resources and enterprises from public to private ownership through the sale of assets previously held by the state. In some cases sales take place through capital markets, with private investors buying shares in former state owned enterprises; in other cases, SOEs are simply sold intact to private bidders. Where possible, state monopolies are broken up in such a way as to allow effective competition. Even without any such break up, it's claimed that the public can still benefit from privatisation.

In response to critics' claims that such privatisation merely transfers 'natural monopolies' from public to private control, neoliberals have referred to Edwin Chadwick's concept of substituting 'competition for a field' for 'competition within a field'. Where competition within a field is impossible, as in the case of local rail networks, for example, the public can still be protected from exploitation, without the need for regulation or public ownership, through auctioning off the franchise to the bidder offering the best price to consumers (Funnell, Jupe and Andrew, 2009, p. 133).

In 1986 in the UK, Margaret Thatcher's Conservative Government sold the state controlled monopoly gas supplier, British Gas, to shareholders, appointing a regulator to control prices and gradually introduce competition (Funnell, Jupe and Andrew, 2009, p. 176). In 1989 the government privatised publicly owned regional water authorities set up in 1974. The 10 authorities were sold off to private businesses for prices significantly below their market value, without any tendering process. The International Monetary Fund (IMF) and the World Bank, which oversee loans to the developing world, made water privatisation a condition of finance for water projects in the developing world (Sjolander, 2005, p. 12). In 1992 a UN water conference endorsed privatisation around the world as a way to improve efficient delivery.

The later Conservative Government of John Major privatised the UK coal industry in 1994, along with British Rail (BR) and British Energy, which operated eight nuclear power plants, in 1996. The BR privatisation involved a 'separation of infrastructure and train operation' with the creation of 25 different franchised train operating companies (Funnell, Jupe and Andrew, 2009, p. 185).

Parallel with these concerns for increased privatisation, neoliberals highlight the need for effective development and protection of intellectual property rights (IPRs), including patents, copyrights and trademarks. Without effective IPRs, there is no incentive for investors to risk their resources in developing new ideas or new products. 'A pharmaceutical company will only have an incentive to invest in the development of new medicines if it enjoys the sole right to the profit on sales of the new medicine' for long enough to recoup potentially very substantial research and development (R&D) costs (Chang and Grabel, 2005, p. 92).

Neoliberals argue that patents and other IPRs were central to the rapid development of the US, British, French and other early industrialising economies (Chang and Grabel, 2005, pp. 92-3). Therefore it is right that the World Trade Organization (WTO), developing laws to regulate international trade, should have strengthened protection of IPRs - with the TRIPS Agreement (*Agreement on Trade-Related Aspects of Intellectual Property Rights*) extending patent life to 20 years - to assist the developing world to follow the same path (Chang and Grabel, 2005, p. 93). Protection of IPRs encourages local innovation and foreign investment and 'makes it easier for developing countries to gain access to advanced technologies and products' (Chang and Grabel, 2005, p. 93). It also encourages businesses in industrialised countries 'to create products and technologies specifically designed for developing countries - such as medicines to fight tropical diseases' (Chang and Grabel, 2005, p. 93).

Fiscal policies

Fiscal policies are government policies relating to revenue raising, through taxation, profit from state controlled enterprise and returns to other government assets, and to government spending decisions and priorities. Decisions are made about the kinds of taxes levied and the rate of such taxes; the scale of payments from (or subsidies to) state enterprise, and spending priorities 'including current expenditure, on salaries of government employees and social security payments and capital expenditures' on infrastructure, including hospitals (Chang and Grabel, 2005, p. 188).

Contemporary liberal right-wingers are generally critical of high levels of government expenditure seeing it as major cause of socio-economic problems. Government spending is seen as inherently wasteful and inefficient because spending decisions are not subject to market discipline and are frequently distorted by narrow political self-interest.

If governments tax highly, to finance high expenditures, then they are taking away funds which could serve to finance efficient private investment and free consumption choices. If governments spend more than they receive in taxes, they produce or increase budget deficits. Increased demand produced by such deficits tends to fuel inflation, which undermines investor confidence.

> More importantly, the government borrowing that is necessitated by budget deficits 'crowds out' private investment. This occurs because the increased demand for loans by the government places upward pressure on the interest rate, which prices many private borrowers out of the market. (Chang and Grabel, 2005, p. 189)

Generous unemployment benefits reduce the incentive to look for work and put upward pressure on wages, thereby threatening profits and continued production. Free access to medical care encourages excess consumption of services. Radical government expenditure reduction is a common component of the reforms required in exchange for IMF assistance (Chang and Grabel, 2005, p. 188).

Friedman versus Keynes

Neoliberal macroeconomics, in its monetarist form, was developed in response to the more social democratic theories of John Maynard Keynes.

J. M. Keynes had argued that the aggregate or total monetarily effective demand in the economy determined the level of investment and production and therefore also the provision of employment. If such demand is too low, businesses have to reduce their output and sack workers. This can ultimately lead to mass unemployment, recession and depression.

Keynes identified various factors determining the level of consumer spending, with various considerations favouring saving over spending and thereby threatening to undermine demand. He argued that those on lower incomes always tend to spend a greater proportion of such income compared to higher income groups so that increasing incomes at the lower level would tend to increase demand by almost as much as the income increase.

He supported both money creation – monetary policy – and government spending and tax cuts – fiscal policy – to reduce unemployment and end depressions. In particular, he called for more government spending on housing, hospitals, schools and transport systems. As business investment is 'driven by fickle animal spirits', it is easy for pessimism about expected returns to cause a reduced level of investment and have a major impact on the economy. Keynes argued that governments should run budget deficits to finance public investment in infrastructure, health and education, where private investment was inadequate to sustain full employment (Pressman, 1999, p. 103). This would drive a positive multiplier of growth with the spending of new incomes creating further demand, further new incomes and so on.

Milton Friedman was highly critical of Keynes' analysis. Friedman 'held that fiscal policy would not work and active monetary policy would worsen the business cycle and lead to greater inflation' (Pressman, 1999, p. 158). He argued that consumption was influenced by expectations about future income, rather than actual levels of current income. So if recipients of increased government spending believed that their additional income would not last, then they would save it, rather than spending it, and thereby undermine such fiscal intervention (Pressman, 1999, p. 158).

Friedman argued that money and monetary policy play the major role in determining economic activity. The key idea here is the quantity theory of money

> which holds that the amount of money in the economy times the number of times each dollar is used in a year to buy goods, must equal economic output sold during the year. (Pressman, 1999, p. 158)

Friedman maintained that while people hold money for reasons other than buying goods, this has little impact on velocity. With velocity relatively stable, 'it is the quantity of money that primarily affects the level of economic activity' (Pressman, 1999, p. 158).

Six to nine months after an excessive increase in the money supply, prices will rise and inflation would become a problem. Too much money creates too much demand and inflation can be controlled only by restricting the money supply. Friedman proposed that the central bank 'be required to increase the supply of money by around 3 to 5% per year – the normal growth rate of the US economy' (Pressman, 1999, p. 158) – to provide the money to purchase additional goods without inflation. He saw the Great Depression as a consequence of bad monetary policy, rather than inadequate demand. The Federal Reserve failed to provide enough new money to sustain economic growth and then increased interest rates when the economy contracted, instead of reducing them, as they should have (Pressman, 1999, p. 158).

Keynes recognised that at full employment, increased money supply, leading to increased demand, would drive inflation. Keynesian economists in the 1960s believed in a stable trade off between inflation and unemployment. Friedman argued that there was a 'natural rate' of unemployment, as an equilibrium level, necessary to maintain a low and stable inflation rate.

Attempts by government to push unemployment below this level through increased money supply would only produce accelerating inflation, with workers demanding pay rises to match expected price increases and employers trying to pass on their increased costs to consumers. This would undermine growth and increase unemployment (producing 'stagflation').

Needless to say, neoliberals have generally gone along with Friedman's critique of Keynes. It seems to justify fiscal inactivity – even in face of significant recession. Such recessions shouldn't occur if the money supply is properly managed. If they do, they are basically self-correcting without external interference. Demand will eventually pick up as a result of naturally falling factor costs and interest rates, and the development of new and more productive technologies.

Neoliberal revenue policy accepts the existence of taxation of corporate profits and individual income in developed and less developed countries. However, it generally rejects taxes on property, including death taxes, on dividend income, on capital gains or on pollution (e.g. carbon taxes). Corporate taxes and taxes on higher incomes need to be kept down to encourage entrepreneurship, investment and economic growth. Progressive income tax is also considered to discourage workers from increasing their productivity. Developing countries might have to depend on international trade taxes – tariffs – because of the underdevelopment of their corporate and income tax collection. They need to move towards accelerated reduction in all such trade taxes to grow economically. All countries should replace high levels of profits, income and luxury sales taxes with flat rate, broadly based consumption taxes which are harder to evade than other taxes.

Central banking – monetary policy

Neoliberals call for monetary policy to be 'non-political', with central banks run by objective and independent economists who can take actions without reference to ruling political groups. Monetary policy is understood here as central bank operations of buying and selling government bonds on the open market or changing the rate of interest they charge the individual banks they lend money to (Chang and Grabel, 2005, p. 180). If banks buy bonds from the central bank this reduces such banks' reserves and therefore the funds they have available to lend. With funds to loan in shorter supply, the cost of such funds – the rate of interest – tends to go up. By increasing the cost of borrowing, this generally reduces investment, spending and growth. When the central bank encourages other banks to sell their bonds, these banks increase reserves and available funds and bring down the interest rate, thereby generally encouraging investment and consumption (Chang and Grabel, 2005, p. 180).

Central banks should be empowered to follow objective policy prescriptions which focus on controlling inflation, rather than bending to political pressures to implement irresponsibly expansionary policies based on low interest rates or printing money to finance government spending. Anti-inflationary policy promotes saving, lending and investment.

> Banks will extend medium and long term loans only if they are confident that their returns will not be undermined by price increases over the lifetime of their loans. Foreign and domestic investment [also] depends on the expectations [of] inflation not undermining returns over the lifetime of the investment project. (Chang and Grabel, 2005, p. 181)

Residents of a particular nation will keep their funds in local banks only if they are confident inflation won't diminish the value of such funds and consumer goods prices will remain stable (Chang and Grabel, 2005, p. 181). If they anticipate inflation

> they will hold their savings outside the country [in the absence of capital controls] and will hoard goods in anticipation of future price increases. (Chang and Grabel, 2005, p. 181)

Free international trade

According to neoliberals the best trade policy is free trade. Free trade allows 'higher rates of output and employment growth', increased 'productivity and efficiency', improved 'living standards and consumption choices' (Chang and Grabel, 2005, p. 55). It reduces the opportunities for corruption and political favouritism.

According to comparative advantage theory, in the absence of government interference, a country will specialise in the production and export of those goods it can most effectively produce 'given its particular endowment of land, labour and capital'. As Chang and Grabel note

> A country has a comparative advantage in an industry if its relative performance in that industry compared to other countries is better than its relative performance in other industries. This implies that every country will have a comparative advantage in something compared to its [potential] trading partners. Even a country that is relatively inefficient in all industries when compared with other countries will nevertheless have a comparative advantage in that industry where its performance is least deficient.
>
> Trade theory contends that under free trade each country can and will specialise in that industry where it has a comparative advantage and will trade with other countries to get those goods for which it does not. (2005, pp. 55–6)

Each country benefits from the exchange of goods and services if, in so doing, it can meet the demand for them from the most cost-effective source.

So by importing some goods, it can concentrate on producing in areas in which it is most competitive, thereby achieving the most efficient and profitable allocation of resources. Through such a process, each country will be better off than it would be if it had continued to produce the goods for which it lacked a comparative advantage, rather than importing them.

The theory provides a basis for rejecting state intervention in production and trade, as such intervention distorts price signals regarding comparative advantage and reduces national productivity and social welfare as a result.

Comparative advantage

David Ricardo originally developed the theory in terms of labour productivity. This can be illustrated by reference to a simple numerical example from Steven Pressman. In country A, output per worker is one manufacture or one tonne of raw material in a year, while in country B output per worker is 3 manufactures or 2 tonnes of the same raw material. Country B has an absolute efficiency advantage in both areas, but a comparative advantage in manufacturing; 3 to 1, as opposed to 2 to 1.

If country A has 200 workers and country B 100 workers equally divided between the two industries of manufacturing and raw materials, then country A will produce 100 manufactures and 100 tonnes of raw material while country B will produce 150 manufactures and 100 tonnes of raw material. The combined output would be 250 manufactures and 200 tonnes of raw material.

However with specialisation, 200 workers from country A, concentrating in the area of least disadvantage, will produce 200 tonnes of raw materials, and 100 workers from country B will produce 300 manufactures, which is an overall gain of 50 manufactures, compared to the self-sufficiency situation. How the output is distributed clearly depends on the exchange rate between the two sorts of goods. If country B exchanges 125 manufactures for 100 tonnes of raw materials from country A, the surplus is shared between them. Country B ends up with 175 manufactures and 100 tonnes, country A with 125 manufactures and 100 tonnes, so that both gain 25 manufactures. With fewer manufactures from country B trading for 100 tonnes of raw materials from country A, the more such trade benefits country B at country A's expense, with more, country A benefits (Pressman, 1999, p. 37).

The theory of comparative advantage suggests that countries are better off removing trade barriers even if their trading partners fail to do so. Their own citizens will still benefit from lower prices for imports and such imports can be inputs for export industries (Chang and Grabel, 2005, p. 57).

It is reluctantly accepted that developing countries may still resort to tariffs to raise government revenue in the absence of other possibilities. So might they need to protect some new domestic industries from foreign competition until such industries are strong enough to survive in the world market. If so, tariffs should be uniform and low (not more than 5%) so as not to distort domestic investment decisions. Industry protection should be limited (with a tariff of not more than 5–10%) and temporary – not more than five to eight years. Such limited tariffs are allowed by the WTO (Chang and Grabel, 2005, p. 58).

International capital flows

Neoliberals are supporters of open international capital markets which give the public and private sectors of particular nation states access to capital and technology not generated domestically. International private capital flows of foreign bank lending, portfolio investment and foreign direct investment increase productive efficiency and 'policy discipline'. The knowledge that foreign investors can relatively easily remove their funding is supposed to provide a strong motivation for political authorities to ensure 'international standards' of macroeconomic policy and 'corporate governance' (Chang and Grabel, 2005, p. 110).

'Liberalisation' of such flows ensures that market forces direct them to the most 'efficient' use (Chang and Grabel, 2005, p. 110). It gives local firms and governments access to the vast pool of capital available on global capital markets. Increased access to such capital can drive positive feedback of increased investment and economic growth. Those in a particular nation state can buy stocks, bonds, derivatives and other financial instruments issued by the private sectors of other states around the world.

Foreign direct investment, as 'the purchase of a controlling interest (10% +) in a business in a country other than the one in which the investor resides' (Chang and Grabel, 2005, pp. 106–7), can provide a country with access to capital and advanced technology, can introduce superior management techniques and business practices and provide links with and access to foreign markets. As Chang and Grabel point out, greenfield investment involves 'the creation of new facilities' providing jobs, tax and technology. Brownfield investment – in the form of 'mergers and acquisitions involving the purchase of assets of existing domestic firms' – can lead to significant efficiency gains (Chang and Grabel, 2005, p. 107).

Since the 1980s the production process has been broken down into numerous specialised tasks or stages, many of which are dispersed around the world. Through foreign direct investment (FDI) countries can benefit from these newly emerging patterns of international production by integration into the global division of labour. The stability of FDI renders it preferable to foreign bank borrowing. The historical and empirical record shows how an open attitude to FDI can promote industrial development, export success and growth.

The resources provided by foreign bank loans, and by loans from the IMF and the World Bank, supplement the pool of capital that is made available by domestic lenders and savers. They thereby provide the opportunity for levels of investment and economic growth that are higher than would otherwise be achievable in the absence of this resource (Chang and Grabel, 2005, p. 106).

Financial deregulation

The neoliberal view favours 'a liberalised financial system based upon competitive capital markets' to ensure 'high levels of savings, investment, foreign capital inflows and economic growth' (Chang and Grabel, 2005, p. 151).

Financial liberalisation is seen as a process in which allocation of financial resources, including provision of credit, for investment, is determined by market forces, with minimal government involvement. As Singh notes, this includes

> ... deregulation of interest rates; removal of credit controls; privatization of government owned banks and financial institutions; liberalization of restrictions on the entry of private sector and/or foreign banks and financial institutions into domestic financial markets; and introduction of market-based instruments of monetary control. (2005, p. 23)

This is sometimes distinguished from capital account liberalisation, which is the process through which countries remove limitations on

> ... domestic banks' foreign borrowing; limiting the entry of foreign capital; and restricting the repatriation of funds from the country. CAL entails dismantling of all barriers on international financial transactions and the purchase and sale of financial or real assets across borders. The IMF and the World Bank have exerted powerful pressures for such financial liberalization in countries that have borrowed from them. Regional treaties, including NAFTA and FCN, as well as OECD codes and EU directives, include obligations to liberalise capital transactions between the nations involved. (Singh, 2005, pp. 23–4)

As Chang and Grabel point out, the proponents of financial liberalisation typically argue that high levels of government financial regulation involve and encourage 'wasteful and corrupt practices', which can be eliminated through the 'discipline' of the market.

> Liberalisation [also] encourages the creation of new financial instruments and markets in which to trade them. This process is termed 'financial innovation'. Investment and financial stability are promoted by these new opportunities for risk diversification and dispersion. (Chang and Grabel, 2005, p. 151)

Indeed there has been a massive proliferation of new instruments under neoliberalism, including derivatives – futures, options and swaps, and securitised mortgage debt. Neoclassical theory views all risk as being of a statistically predictable character. It sees the continuing development of financial markets as the best way for the economy to cope with risk, allowing investors to 'hedge' against losses in increasingly sophisticated ways. With diverse types of investments, different sorts of risks balance out – gains cancel losses. With lots of innovations, market forces allow proliferation of the good ones, expanding the health and growth of the system.

Using an example from Eatwell and Taylor, we can see that when a farmer sells their grain 'forward' at a specific date for a given price – as a 'futures' derivative – they are insulating themselves against price fluctuations. The buyers of such a derivative are

> ... taking on the farmer's risk. If, when the harvest was in, the actual price was lower than agreed, they would still be forced to buy the corn at the previously contracted high price. If the actual price on the day the contract closed was higher, they would make a profit. The risk did not disappear, it was assumed by those willing to shoulder risk, for a fee, and they would offset their exposure to risk by spreading it through another range of transactions. (2000, p. 100)

As Eatwell and Taylor point out, such agreements provide 'insurance' through diversification of risk and loss sharing 'by a broad section of society'.

Liberalisation is said to increase the availability of finance and allow borrowers to choose from a wide range of different options, what best suits their requirements without resorting to 'informal financial arrangements' (Chang and Grabel, 2005, p. 151). The finance provided by stock and bond markets is regarded as superior to other forms of finance because it allows individuals and institutions to have diversified portfolios of investments, thereby spreading risk. Such allocation is based on profitability and therefore upon efficiency. It generally costs less than bank funding, and it is 'highly liquid', with individuals and institutions more willing to invest because they can sell their shares and bonds whenever they need to (Chang and Grabel, 2005, p. 151).

As Chang and Grabel note, proponents here emphasise the 'disciplinary' role of such liquidity with investors abandoning underachievers, and the way in which 'internationally integrated capital markets' assist in the integration of developing nations into 'the global financial system' (Chang and Grabel, 2005, p. 152).

By contrast, regulated systems are characterised by high levels of 'state involvement and the domination of banks rather than capital markets' (Chang and Grabel, 2005, p. 150). With the state maintaining artificially low interest rates to encourage borrowing for investment, domestic savers are motivated to hold their funds abroad or use them for domestic consumption. This gives domestic banks inadequate funds to extend investment loans.

Regulation

While generally favouring deregulation, neoliberalism has supported some basic regulatory institutions and processes. In Australia, these have included the Australian Competition and Consumer Commission (ACCC) and Australian Securities and Investments Commission (ASIC). The ACCC is supposed to play a central role in enforcing anti-monopoly and consumer protection legislation. ASIC is concerned with investigating and prosecuting suspected contraventions of the laws governing the operation of corporations, including company or financial services fraud or dishonesty. It is concerned with consumer protection in financial services, the regulation of insurance and superannuation, and monitoring and promoting market integrity of the Australian financial system.

The operations of the ACCC are directed and justified by the basic neoclassical idea of equilibrium in a freely competitive market. Supposedly, state regulators maintain an appropriate equilibrium of supply and demand in each market which is optimally efficient in providing customers with quality products at the lowest possible prices. This is achieved through breaking up price-fixing cartels and restricting mergers and takeovers that threaten to create concentrated corporate monopoly power.

The operations of ASIC and of other relevant bodies in the regulation of financial markets are also directed and justified by an idea of 'efficient stock markets' similarly centred upon a model of rational and efficient market equilibrium. Orthodox economists tell us that share market valuations generally reflect the true future prospects of companies. The efficient markets hypothesis claims that the collective expectations of stock market investors are accurate predictions of the future prospects of companies; that share prices fully reflect all information relevant to the future prospects of the traded companies and that changes in share prices reflect changes in information relevant to such future prospect, which arrives in random fashion (Keen, 2001, pp. 226–7).

Contemporary ideas and practices of regulation are built around these ideas – of the true values as equilibrium point around which stock prices oscillate in response to random fluctuations in the supply of information.

Such regulation is supposed to ensure that investors are properly informed about the companies whose shares they hold, thereby keeping market prices close to equilibrium prices, as well as managing 'systemic risk' by preventing 'market abuse' and economic crime.

The assumption is that investors tend to take excessive risks which pose a threat to the whole economy. Capital adequacy standards require banks and other financial institutions to hold liquid capital in proportion to the extent of their investment risks.

Free currency markets

Neoliberals generally favour what are called 'floating exchange rate' systems with full currency convertibility. This means that anyone is free to exchange any amount of any currency for any other currency. The relative values of all currencies are determined by market forces and central banks are committed 'to buy or sell unlimited amounts of the domestic currencies' at the going market price to whoever wants them (Chang and Grabel, 2005, p. 164). Increase in demand for a particular currency causes its value to appreciate relative to others. This will happen, for example, when foreign investors purchase assets in the country, because they have to obtain the currency in order to pay for such assets.

A reduced demand, brought about when investors sell assets denominated in that currency and then sell their holdings of that currency, leads to depreciation. This contrasts with a 'fixed or pegged' system, where 'the value of the currency is set by the government or allowed to fluctuate within a narrow band' (Chang and Grabel, 2005, p. 165) and with non-convertible currencies, where central banks and other money holders within a state can sell currencies only 'for pre-approved types of activities following purchase of a foreign exchange licence' (Chang and Grabel, 2005, p. 164).

Neoliberals argue that full convertibility is crucial to encourage foreign investment and avoid corruption with resources wasted in trying to avoid convertibility controls. Floating rates allow market forces to ensure maximum economic efficiency and balanced international trade. Increasing currency values signal the confidence of investors in a particular economy and thereby encourage further investment of funds where they can be best used. They increase the costs of the country's exports thereby reducing trade surpluses, but they also reduce the country's import costs, thereby reducing the cost of exports with a high component of imported materials. Decreased values increase import costs to the country in question, threatening balance of payment problems. If A$1 exchanged for US$2 yesterday but only US$1 today, then imports from the US to Australia will be twice as expensive in Australian dollars today as yesterday. Such devaluation signals the need for reforms, while also reducing the costs of investment in, and exports from, that country, thereby offering assistance in addressing the problems through increased exports and decreased imports.

Floating rates promote a greater degree of financial stability than do fixed or pegged rates, since the latter encourage speculators to push governments to force a change in the rates, and thereby cause a collapse of such a pegged regime.

Neoliberals acknowledge that particularly weak and vulnerable economies may not be able to withstand the currency fluctuations produced by floating rates. In this case they call for rigidly fixed exchange rate regimes based on a 'currency board' or full currency substitution. Such a board is charged with holding sufficient reserves of a stable overseas currency (typically US dollars or Euros) to ensure convertibility of the local currency

> The... board can only issue additional domestic currency if holdings of the reserve currency are increased by export sales or foreign investment inflows... Full substitution involves legal replacement of the domestic currency [by] a strong foreign currency. (Chang and Grabel, 2005, p. 167)

Currency boards or substitutions are seen to maintain currency stability and prevent high inflation through excess domestic money supply and thereby encourage foreign investment (Chang and Grabel, 2005, pp. 167–8).

The IMF, World Bank and the WTO

The Bretton Woods Institutions of International Governance, the IMF and the World Bank, created by the US and UK in 1944 to oversee postwar economic development, have been transformed from being supporters of a highly regulated international monetary system, of capital controls and fixed exchange rates, to being the principal drivers of increasing deregulation. Neoliberal ideas have also played a key role in the development of the WTO as a force similarly committed to such deregulation and privatisation.

In particular, the politically appointed executive directors of the IMF and the World Bank have taken advantage of the debt crisis created by massive World Bank and private bank lending to developing countries from 1968, followed by big US interest rate hikes in 1981, to impose rigorous 'structural adjustment programs' upon the debtor nations as conditions for renegotiation of such debts and access to further loans.

These programs have included such key features of neoliberal policy as: encouragement to competition at all levels of society; inflation minimisation; concentration on exports and increased trade volume, typically achieved through devaluation of the local currency to increase exports to provide foreign currency for debt service and modernisation; removal of all restrictions on foreign capital flow into the countries in question; reduced taxation of corporations and rich individuals to encourage such investment; privatisation of any government owned resources and introduction of cost recovery or user pays for any previously free or subsidised goods or services; and the introduction of 'flexible' labour markets through removal of workers' rights to job security, health insurance, maternity leave or minimum wages.

In line with neoliberal policy, as outlined earlier, the IMF Structural Adjustment Programs (SAPs) require the creation of independent central banks and/or constraints on the operation of central banks, precluding money creation to finance government budget deficits, and inflation targeting to keep inflation below a predetermined level, usually 2–3% (Chang and Grabel, 2005, p. 181). Supposedly, such 'Washington Consensus' policies, if pursued actively enough, will encourage growth and eliminate debt.

The WTO developed out of a series of international negotiations – the General Agreement on Tariffs and Trade (GATT) rounds – aiming to reduce tariffs and increase world trade. It functions to oversee the implementation of a number of agreements signed in Morocco in 1994, following 8 years of the Uruguay Round of negotiations. As Susan George says

> these agreements cover not just [trade in] industrial goods, but agriculture, services, intellectual property, technical standards and a Dispute Resolution Body whose decisions are binding. (2004, p. 58)

The basic intent of all of these rules for the regulation of international trade and development is to 'achieve a progressively higher level of liberalisation' including privatisation of government services, and unrestricted foreign investment in all economies. Any ethical 'restriction' upon trade or investment based upon 'processes or methods of production' – including wages and conditions, child labour or environmental vandalism – is forbidden. A major argument here is that such ethical discrimination would be disastrous to the developing world, where economies need to grow – through trade and investment – in order to be able to afford protection of labour and the environment.

Ethical issues

Neoliberals frequently follow many neoclassical theorists in presenting their favoured economic models and policies as 'objective' science which is far removed from 'subjective' ethical speculation.

If we want the economy of our nation and of the world to work 'efficiently', by making the best use of available resources to best satisfy consumer demand, and achieving the maximum possible level of productive output without wasting available resources, as measured by gross domestic product (GDP) per capita, then these are policies we must pursue. These are the policies which any truly 'responsible' or 'rational' government would follow.

If the economy is functioning efficiently, it is providing the material for ethical and political decisions about welfare and justice. If resources are wasted through inefficiency, then this reduces the capacity for any such decision making.

Beyond this point, there is a range of different responses to issues of ethical justification of free markets which appeals to neoliberal thinkers. These responses fall into three main categories: the prioritisation of liberty and property over issues of justice; utilitarian or Pareto-based defence of free market relations; and particular versions of just deserts.

As considered in Chapter 1, the prioritisation of liberty over justice goes back to the classical liberal philosopher John Locke, who argued the case for inalienable human rights of legitimate acquisition and ownership of private property.

In so doing, he provided an ethical defence of free capitalist market relations without reference to utility or to just deserts. Robert Nozick argued the case that real freedom in the exercise of property rights is fundamentally incompatible with any kind of patterned justice principle such as equality or just deserts, which could be achieved only through ongoing totalitarian state intervention into individual life.

Free markets inevitably produce inequality. But insofar as such markets are genuinely free in the sense that individuals participate without coercion or fraudulent misrepresentation then so do such participants freely choose the consequences in question. They themselves are responsible for any inequality which results.

Many neoliberals also go along with key ideas of neoclassical welfare economics in arguing that efficient markets have intrinsic tendencies to distribute goods in such a way as to maximise – objectively measurable – social welfare and achieve genuine social justice. On the distribution side, there is the idea that free market forces generate Pareto-optimal outcomes 'where it is impossible to make anyone better off without making someone else worse off' (Hahnel, 2002, p. 32).

Given competitive conditions, rational consumers and rational entrepreneurs will automatically act and interact – through exchange – to maximise their own well-being. Consumers weigh up prices of goods against their costs to make the most efficient use of their resources in 'maximising their utility'. Competition and business efficiency ensure that consumers acquire products at the lowest possible cost. At the same time, through such efficient organisation of the production process, and through producing goods up to the point where 'marginal cost equals marginal revenue', firms are able to maximise their profits.

Since individuals would not exchange goods unless both saw their situation improved by such exchange, ongoing exchange leads to ongoing improvement.

Some utilitarians and some human rights theorists claim to be able to provide objective measures of individual and social welfare in terms of levels of mental and physical health and well-being and realisation of human potentials. Neoliberals dismiss this as arrogant and authoritarian. Individuals know, better than others, what is good for them and how much good they derive from particular items of consumption. This is reflected in the price they are prepared to pay for such items.

Neoliberal appeals to 'just deserts' are generally intertwined with utilitarian defence of free markets. As noted in Chapter 1, the key idea here is that of reward for the value of productive contribution, and/or for the amount of effort, sacrifice or risk involved in valuable productive contribution.

As Robin Hahnel points out, just deserts can be taken to refer to the productive contribution of human capital of individual strength, skill and knowledge, or of physical capital of an individual's money or material resources. Similarly, effort and sacrifice can include both individual contribution of unpleasant or unrewarding labour, or the sacrifice and/or risk involved in investing funds in production rather than spending them on immediate consumption (Hahnel, 2002, pp. 24–31).

The suggestion here is that greater effort, sacrifice or risk is typically needed to create greater value, through the use of both human capital and physical capital. The free market generally rewards the creation of greater value with proportionally greater monetary payment. So free markets are generally fair and just in appropriately rewarding effort, sacrifice or risk.

In utilitarian terms, ideas of reward for productive contribution of value, effort or risk, are bound up with the idea of a requirement for appropriate incentives to motivate necessary but unappealing productive activity or provision of capital. Such incentives are required in order to motivate essentially selfish – but rational – humans to contribute to the general social good. Through rewarding labour in proportion to the value of output, markets provide effective motivation necessary to drive useful production.

Key value creating roles require unrewarding and expensive training and involve high levels of stress. Substantial rewards are, therefore, required to motivate those able to develop and exercise such skills to do so. As few have the necessary drive or ambition or the innate talents which can be trained into the skills in question, such skills will always be in short supply and the market, therefore, ensures high returns to such individuals.

When someone makes physical capital available for productive purposes, this involves a sacrifice on their part of the immediate consumption that could be pursued through sale of such resources. In terms of just deserts, this requires a proportionate reward. Radically new technologies and new products can drive the whole society forwards with huge benefits for all – but they can also spectacularly fail. The risks of investment in such innovations can be great. Those prepared to take such risks, to gamble on such significant progress, deserve proportionally greater benefits where they succeed. In utilitarian terms, people would not continue to take such risks – and society would be the poorer – if major losses were not offset by proportionally greater than normal gains.

Some might say that wealth inequalities undermine the equality of opportunity for younger generations, especially with private provision of education and health services. If some start out with greater wealth generating capacities than others, by virtue of superior education, for example, then it is unfair that everyone should be rewarded in proportion to the wealth that they create.

Neoliberal supporters of 'just deserts' typically argue that high levels of social mobility, supported by anti-discrimination legislation, offset such initial inequality of opportunity – allowing all who choose to take appropriate steps of skills acquisition and exercise to be able to radically increase their value creating capacities and hence rewards.

Where neoliberals do acknowledge limits to such social mobility, they argue that any attempt to further reduce such inequality would have disastrous consequences for society as a whole. As Hayek said, government intervention to try to achieve this will only stifle economic growth and efficiency, to the detriment of all.

DISCUSSION TOPICS

1. Are there any practical or ethical problems with the orthodox economic ideas and interpretations favoured by neoliberals, in relation to the following?
    free markets
    corporations
    independent central banks
    free international trade
    financial deregulation
    privatisation

2. GDP is the sum of all final expenditures on goods and services within an economy, comprising consumer spending, government spending, gross investment in fixed capital and net exports. Is maximisation of GDP and GDP growth (per capita) really a good index of the economic health of a society, as suggested by neoliberal economics?

3 Consider the failure of GDP to address distribution issues, environmental costs of pollution and resource depletion, non-monetary values, depreciation of capital stocks and debt. Costs generated by avoidable social problems are counted towards health and growth. What are the ecological limits to GDP growth?

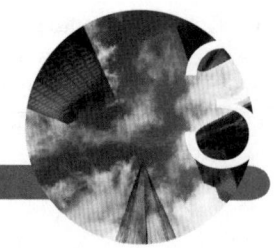

Problems of competition

This chapter begins to consider some of the theoretical and practical problems associated with 'orthodox' economic ideas. Poor people in the developing world have been the major victims of such ideas, with nearly a billion people left without the basic necessities of life, and the numbers are increasing every year. In the US, the working class have also suffered significantly, struggling for longer hours to earn less in real terms in 2009 than they did in 1979. The poorer fifth of world population receive around 2% of global income, while the richest 20% receive three quarters of it.

Three decades of neoliberal economic policies have culminated in the biggest world recession since the 1930s, with more than half the entire value of tradable shares everywhere wiped out in less than a year. Around a quarter of the world's financial wealth has been destroyed. A meltdown on Wall Street has been followed by a global banking crisis and multi-billion dollar government bailouts around the world, with even the most conservative pro-marketeers acknowledging that this is a result of hopelessly inadequate regulation of financial markets. Subsequent 'stimulus packages' have hugely increased government debts. Meanwhile, the output of greenhouse gases continues to increase, threatening catastrophic climate change – including destruction of food and water supplies – in the not very distant future.

In many ways, neoliberal economics can be seen as the ideology of monopoly and oligopoly capital, driving political program that optimally favour owners and controllers of such big businesses. The emphasis upon free competition is misleading because unregulated competition itself generates concentrated monopoly power.

Monopoly and oligopoly

The original proponent of free market efficiency, Adam Smith, acknowledged the difficulties of achieving 'free competition' in practice given the very strong interest of business people in reducing or eliminating such competition. So is it clear that processes inherent within such free competition also work to undermine it in the longer term.

Free market competition tends to generate uncertainty and instability, particularly where the actual or potential number of competing suppliers of a given product or service is large. While producers of a particular good can gauge the size of market demand through market research, it is much more difficult for them to estimate future supply of the good in question (Shutt, 2001, pp. 47–8). They do not know when their competitors will apply new cost-cutting techniques and technology. In face of such uncertainty they seek to protect themselves from competition by pursuing cost-cutting technologies and economies of scale themselves. This leads to an expansion in overall productive capacity beyond a level compatible with available demand, leading in turn to sharp cyclical fluctuations in output, employment, profit and price.

Periodic downturns create high levels of bad debt, putting huge pressures on the financial system, threatening mass failure of the banking system, leading to collapse of the productive economy. Such disruption creates problems for owners and controllers of capital, as some businesses collapse and others struggle. The primary victims are working people without property or savings to cushion them when they lose their jobs. Without adequate welfare provision, unemployed workers are threatened with destitution and disaster.

Such fluctuations produce an increasing concentration of ownership in a diminishing number of enterprises as weaker, typically smaller, higher cost, competitors are forced out of business. Large firms typically have cost advantages over small ones because of scale economies and, given open competition, they drive smaller ones out of business – if they want to (Shutt, 2001, pp. 48–9).

In the past, the regular periodicity of the business cycle appears to have been associated with the turnover time of major components of fixed capital. As new, more productive technology becomes available, the first to innovate can make super-profits, selling at a price below that determined by the old production costs. They can increase the scale of their productive operations by taking market share away from others. This forces others to catch up, and as the technology is generalised the sale of goods price falls – because of competitive price reductions and increasing oversupply – in line with the new, reduced costs. The new technology has to be paid for and the latecomers have less revenue to finance this. For them the new technology eats into profits, rather than boosting them (Harman, 1984, pp. 14–49).

Orthodox theory says that the survivors are those who produce the best deal for consumers in terms of quality and price. In practice they tend to be the biggest and strongest with resources that allow them to survive a competitive price squeeze. The larger operations typically benefit from returns to scale. Their power is also increasingly used to prevent new competitors entering the market, by means of predatory pricing, pressure on distributors, or by buying out competitors through purchase of their shares. There is, therefore, an inevitable tendency for competition to be reduced as increasingly big and powerful companies seek to protect their investors' return to the disadvantage of consumers (Shutt, 2001, p. 48).

These increasingly large corporations operate effective cartels or informal price fixing arrangements, reducing supply to keep prices high. Such developments are supposedly addressed by regulatory bodies enforcing anti-monopoly legislation in the developed world. However, politically influential mega-corporations have the power to water down the rules and their application, as far as they themselves are concerned. They are able to effectively pressure governments, which are reliant upon them for jobs and tax revenues.

Monopolies themselves become increasingly constrained by the monopolisation process, insofar as they have to pay each other's monopoly prices for necessary inputs to their production processes, so that their higher than average profits are dependent on the size of the non-monopoly sector. With restricted production reducing job opportunities and wages, they face problems of inadequate demand for their products without a huge expansion of easy credit.

Advocates of free markets argue that such markets offer 'freedom of choice' to consumers whose rational purchasing decisions guide the production process. But monopoly price-fixing and market sharing radically restrict the choices available to consumers, while massive advertising aims to undermine their capacities for rational choice, enabling corporations to charge high prices for shoddy and unnecessary goods. Here again, consumer protection legislation supposedly functions to address these issues in the developed economies. But governments and regulatory bodies remain susceptible to corporate pressure in relation to the formulation and application of the rules.

Undermining democracy

The inevitable subversion and subjugation of the democratic process is a major social problem of increasingly concentrated private economic power.

The vast power of big corporations and super-rich individuals to acquire media monopolies to shape public opinion, to finance multi-million dollar election campaigns of their chosen parties and candidates, to bribe officials at all levels, and to blackmail incumbent regimes through threats of capital flight and disinvestment, radically undermines meaningful political choice for the majority of the population. Such corporations dictate legislation and direct increasing quantities of taxpayers' money into their coffers as corporate welfare.

Support for private property rights of an increasingly wealthy minority of corporate leaders inevitably undermines basic democratic rights to freedom of speech, organisation and campaigning. This is because democracy depends upon genuinely equal access to such rights by all citizens. A minority monopoly of private wealth can be turned into a minority monopoly control of the political process through control of the media, and funding of campaigning and of government. Such effective disempowerment of the majority leads to increasing disillusion and alienation from the political process for increasing numbers of ordinary people – which only serves to further increase the effective political power of private wealth.

Domination of the political process by the elite leadership of a couple of well-funded 'mainstream' parties, particularly in the English-speaking world, has become the means for the exercise of real power by, and in the interests of, the principal financial backers of such major parties. As Miliband (1994, p. 73) notes, 'institutional independence masks the very real dependence of legislatures on the "special interests" represented by business'. So that even where the legislature appears independent from executive authority and free from 'party discipline', as with the US Congress at election time, and other times, such senators and representatives remain crucially dependent on, and therefore beholden to, the support of the same corporate business interests and lobbies that fund campaigns for such executive authority. In the US, election funding of the major political parties by large private sector corporations is 'estimated to have reached around US$3 billion in the federal election of 2000' (Shutt, 2001, p. 62).

Such control of election funding and the mass media is supplemented by increasing corporate control of education and research. The power of big business can also corrupt the political process through promises of lucrative sinecures for retiring politicians and the threat of financial sabotage through disinvestment, capital flight and job loss, and through politicians' and their families' corporate shareholdings (Shutt, 2001, p. 63).

Major political parties around the world have been effectively bought off by big business. This is reflected in a lack of any real policy differences between them. Up until very recently all subscribed to basic elements of the neoliberal consensus with no substantial disagreements in fundamental policy areas, and collaborated in preventing serious consideration of alternatives. The scale of the current crisis is so great that it has begun to undermine such orthodoxy in some areas. It remains to be seen whether any real alternatives will have significant effect upon public policy.

We must also consider the even greater limitations of many of the new democracies in the developing world (and in parts of Asia and Eastern Europe), where many governments have appalling records of repression and abuse of basic human rights of their own citizens. Elections in some nations involve high levels of bribery, corruption and intimidation, and restriction of opposition parties campaigning in some or all areas. These political problems can be traced back, to a great extent, to the economic weaknesses and disadvantages of such nations, oppressed, victimised and manipulated by more powerful nations – and big corporations – in the global marketplace.

Innovation, sustainability and waste

Supposedly, free markets drive rapid and ongoing productive innovation in the service of competition. But increasingly concentrated private monopoly power typically has a strong interest in – and capacity for – suppressing new technology that threatens to devalue its current productive resources. Defenders of 'free markets' for goods and capital argue that market forces ensure the speedy development and application of the most efficient technology and that this increasingly means the greenest, least polluting and safest technology. However, the development of genuinely sustainable technologies involves time and resources. Wherever a business shifts funds away from shareholders' dividends and short-term profit maximisation into longer-term programs; it is likely to be severely punished by the share markets.

At the same time, increasing concentration and centralisation has given big monopoly corporations the power to suppress the development and application of sustainable technologies. They have achieved this by taking over new businesses that are seeking to develop and apply such technologies or by destroying such businesses through temporary price reductions for their own output. They pressure governments to penalise rather than support any newcomers. The effective life of older, unsustainable technologies – and consequent pollution and environmental destruction – is thus extended.

Competitive markets are intrinsically wasteful and destructive. They depend upon initial duplication of productive resources – often built up at great cost – only to see such resources discarded, as particular businesses, regions and nations fail in the competitive struggle, with workers losing their jobs. Substantially reduced demand then contributes to further failures. Insofar as economies of scale are supposed to win in the end, questions arise of why such economies were not planned in the first place (rather than going through crisis and recession in the process of producing them) and why they should not be made to serve the public interest, rather than monopoly private profit. These considerations are particularly significant to sectors of the economy requiring substantial investment in infrastructure networks, such as the roads, railways, water supplies, energy distribution and telecommunications. Here even orthodox economic theory recognises the inefficiency of creating competing networks, with wasteful duplication of expensive infrastructure to provide services that could be achieved with a single network. This has not prevented such wastage of resources in ideologically driven neoliberal privatisation campaigns. Ecological concerns now make rational central planning of infrastructure development absolutely vital.

Public versus private monopoly

So, far from establishing a balanced equilibrium of lowest possible cost to consumers, a perfectly competitive market is inherently unstable and develops into oligopoly, with several large firms fixing prices in an industry, or monopoly.

It's true that the ideal scale of output can be different for different industries. In some cases, smaller-scale production can be more efficient and productive (there can be diseconomies of scale). But not surprisingly, as a result of market competition, most industries have come to be dominated by a small number of very large firms. This includes areas of oil production and distribution, steel and aluminium production, car and computer manufacture, food processing, production of aircraft and of laundry detergents, airline travel, and banks (Stillwell, 2002, p. 180).

Competition between smaller firms typically precludes the productivity benefits of larger-scale productive organisations and obstructs large-scale planning and integration of productive operations in the service of rationally – and democratically – chosen goals. The collapse of smaller weaker businesses, failing to compete with bigger ones, involves a significant waste of resources with the miseries and consumption disruptions of unemployment. It makes a lot more sense, in most cases, to start out with significant scale benefits and regulate monopoly prices.

This is precisely what has happened when countries have needed to be genuinely economically efficient – to fight in wars that threaten their total destruction or drag themselves out of rural poverty and backwardness to become developed economic powers in a hostile and competitive environment. State power has presided over the creation and operation of productive monopolies in a context of high levels of regulation of trade and investment.

As Ha-Joon Chang (2007, p. 59) points out, Japan's unprecedented post-war industrial development was driven by the central planning of the

Ministry of International Trade and Industry (MITI), with tight restriction of imports through foreign exchange controls, and direct and indirect control of exports, banning of all foreign investment in key industries, and the channelling of subsidised credit to key sectors. The government presided over the creation of waste-reducing monopolies. Similarly, in South Korea, the government maintained strong tariff protection of national production, restricted and regulated foreign investment, retained control of all of the banks, major industrial operations and foreign exchange, channelling credit and subsidies to key sectors, in the national interest (Chang, 2007, p. 14).

Strong central state control allows for production in the service of particular social and political priorities, rather than merely of private profit. Such priorities can be more or less enlightened, more or less directed to the protection of human rights and the fulfilment of human needs and protection of the environment.

Problems of privatisation

The neoliberal privatisations, referred to in Chapter 2, were good for those able to acquire valuable assets at budget prices, less good for taxpayers and users of the services in question, particularly poorer people. These privatisations resulted in deterioration in infrastructure, declining health and safety provision for workers and the public, increased prices to consumers, and diminished government capacity for stabilising the economy (through maintaining jobs in downturns) (Funnell, Jupe and Andrew, 2009, p. 127). Large numbers of workers lost their jobs without finding employment elsewhere. In most cases, the public ended up paying more both in terms of government subsidies and government regulation, as well as in direct costs for the services in question, than had been the case with previously fully controlled government instrumentalities.

Such privatisations involved direct transfer of monopoly power from public to private control – as with British Gas and British Telecom in the UK – with monopoly profits flowing into the pockets of wealthy shareholders and executives and away from subsidies to more vulnerable consumers and state welfare provision. Or they promoted the breakdown of efficiency and safety through fragmentation and duplication of services.

As Funnell, Jupe and Andrew (2009, p. 126, pp. 128–50) point out, 'transport, in particular, is a capital intensive industry which requires long term investment programs. It may be possible to operate some parts of the transport infrastructure on a commercial basis, but not all can generate sufficient revenue to cover their full costs.' The breakdown of British Rail, to create a fragmented, complex and poorly organised system of private operators, led to loss of scale economies, burgeoning costs and disintegration of infrastructure and safety standards.

Monetary policy

Free marketers have always called for low tax and low government spending. They have looked to monetary policy, controlled by 'independent' central banks, rather than fiscal intervention, to ensure the harmonious growth of the system. On the one hand, this has meant a tightening of monetary policy if and when accelerated growth has threatened to improve the bargaining position of labour through approaching genuinely full employment. On the other, it has meant a loosening of such policy whenever the unearned – and in some places untaxed – investment incomes of the rich are threatened by major share market readjustments.

In the earlier stages of neoliberal domination, big interest rate increases, justified as necessary to fight inflation, served to boost unemployment and weaken workers' organisations. Key developments in the swing to neoliberal policy were the election of Margaret Thatcher as British Prime Minister in 1979, with 'a mandate to curb union power' and to put an end to stagflation, the takeover of the US Federal Reserve by Paul Volker later that same year, with full employment policies sacrificed to the 'fight against inflation' and the election of Ronald Reagan as US president in 1980, similarly committed to deregulation of markets and destruction of the power of organised labour.

These 'reformers' acted swiftly to push up interest rates in order to destroy less efficient smaller businesses and increase unemployment. The aim was to fight inflation by reducing average production costs – most importantly workers' wages and conditions. Thereafter, the neoliberal leaders supported supposedly 'independent' central banks, at arm's length from the government and committed (solely) to 'inflation control', through interest rate manipulation, rather than full employment through balanced monetary and fiscal interventions. Indeed, the role of such banks was to push up interest rates if demand for labour threatened to increase its cost, or if such demand threatened to allow trade unions to exercise any significant power ever again.

It's not true that neoliberal macroeconomic policies (of inflation control) have supported some objective 'common good' of economic growth, as argued by their proponents. While inflation control through high interest rates benefits the financial community, including those with money to lend and money to deposit, industrial producers and exporters suffer from the increased cost of investment and appreciation of domestic currency produced by such high interest rates. It's more difficult for exporters to sell their now more costly goods overseas. Increased interest rates in the economic growth phase of the business cycle deny workers jobs and pay increases. The independence of central banks sits badly with a supposed commitment to democracy, especially considering the profound significance of monetary policy for everyone (Chang and Grabel, 2005, p. 183).

Empirical evidence shows no clear correlation between central banks' independence and anti-inflationary outcomes in developing countries. It's not associated with higher growth rates, employment, financial stability, balanced budgets or avoidance of fiscal deficit monetisation. While neoliberals argue that central bank independence is necessary to attract foreign capital flows, foreign investors have been quite willing to invest in countries that don't have independent central banks like Russia and China (Chang and Grabel, 2005, p. 184).

As Chang and Grabel (2005, pp. 184–5) note, the central banks of Japan and Europe played a leading role in their industrialisation, directing 'subsidized credit to strategic sectors of the economy as part of industrial policy programmes' and organising 'the distribution and price of credit allocated by the banking system'. Only in the 1990s with neoliberal ascendency, did things change.

The fight against inflation can have disastrous consequences for industrial activity, employment and economic growth. On the other hand

> Numerous empirical studies suggest that moderate levels of inflation have little or no cost in terms of economic growth with economic costs only at high levels. Japan and Korea... grew rapidly in the 1960s and 1970s with relatively high inflation [of] around 20%. (Chang and Grabel, 2005, pp. 184–5)

Taxation, welfare and wages

The class basis of neoliberal policies is clear in relation to taxation, wages and welfare. Where strongly neoliberal regimes have gained power, particularly in the English-speaking world, they have radically cut back social welfare spending for the poor and disadvantaged – even as high interest rates have boosted unemployment – while offering tax cuts to the rich. They have failed to take effective action in many areas of tax avoidance and tax minimisation by wealthy elites. They have failed to take action to protect workers' pay and conditions or to rein in burgeoning executive salaries.

As Wilkinson and Pickett observe

> In 2007 chief executives of the largest US companies received well over 500 times the pay of their average employees, and these differences were getting bigger. In many of the top companies the CEO is paid more in each day than the average worker is in a year. Amongst the Fortune 500 companies the pay gap in 2007 was close to ten times as big as it was in 1980, when the long rise in income inequality was just beginning. (2009, pp. 242–3)

Little of this inflated executive remuneration is given back to governments in the form of progressive income taxes. In 2007, the maximum tax rate, for incomes over US$350 000, was 35%. Around the world, the wealthy have ways of avoiding tax, not available to poorer people, through use of tax havens, family trusts and transfer pricing.

The ratio of executive to shop floor remuneration is a rough index of the extent of neoliberal domination of ideology and politics. Looking at all the companies in the manufacturing sectors of selected industrial nations, Wilkinson and Pickett (2009, p. 243) estimate the ratios of chief executive officer (CEO) compensation to production workers' pay in 2009 to be 16:1 in Japan, 21:1 in Sweden, 31:1 in the UK and 44:1 in the US. They refer to International Labour Organization (ILO) research showing no relation between executive pay and company performance. 'Excessive salaries' rather 'reflect the dominant bargaining position of executives.'

In Australia the dividend imputation credit system has increased the proportion of income going to the wealthy minority of major shareholders.

> Ostensibly designed to alleviate double taxation of dividends, this imputation system allows tax paid by companies to be refunded to shareholders receiving franked tax-free dividends. The effect has been a massive redistribution of income to corporate shareholders. (Stillwell and Jordan, 2002, p. 155)

Wages have become increasingly uneven across occupations and within them, with a much greater proportion of managers and administrators in the top income brackets. Stillwell and Jordan show that the ratio of CEO income to average earnings increased in Australia from 18:1 in 1989–90 to 63:1 in 2004–05, with one major bank CEO, for example, receiving 'an annual pay of A$21.4 million in 2005', equivalent to A$400 000 per week.

The federal Labor Government started the process of dismantling collective bargaining and centralised wage fixing in the 1980s. The Howard Government's *Work Choices* legislation of 2005 radically reduced the range of safeguards and benefits provided by previous awards, further 'undermining the collective power of organized labour, making it possible for employers to offer contracts to individual workers on a take-it-or-leave-it basis', leading to 'further redistribution of income from labour to capital and further widening of income disparities'.

While the staunchly neoliberal Howard Liberal-National Coalition Government 'significantly reduced welfare payments to some of the most disadvantaged groups in society', so too did they deliver substantially higher tax cuts to those on higher incomes, with 'the poorest 50% of taxpayers receiving just 19% of the cuts' in 2005–06. Big cuts in income tax for the wealthy were complemented by 'cuts in company tax rates... and reduction in the effective rate of capital gains tax'. Following the 'effective halving of that tax in 1999, 68 000 taxpayers earning over A$100 000, less than 1% of taxpayers, received half of all the capital gains received by individuals during that year... amounting to an average tax cut of A$220 per week' (Stillwell and Jordan, 2002, p. 155).

The developing world

From the later nineteenth century, big monopoly corporations in the West directed government policies of imperial domination of the less developed world, destroying indigenous industries to turn the colonial territories into suppliers of cheap raw materials to the colonisers' home industries and markets for their industrial manufactures.

When such colonies achieved independence, in many cases in the decades after the Second World War, they suffered major disadvantage due to the lack of infrastructure to support modern sovereign states. They remained dependent upon aid, loans, investment and markets provided by the original colonisers, who continued to exercise control without the responsibilities of direct rule. Intense competition between developing nations for such developed world markets brought down the prices of their (in many cases substitutable) raw material exports, while they were forced to pay monopoly prices for Western manufactures.

Developing nations were allowed limited import restrictions and tariffs to protect infant industries and provide government funding for infrastructure while exporting raw materials to raise funds for industrial imports.

Ownership of land and mineral resources by foreign corporations and local feudal elites continued to contribute to widespread poverty and privation in many regions. This motivated peasant movements and reforming governments to try to nationalise such resources and redistribute productive land to landless peasants. This in turn led to military responses by the (now 'neo') colonial powers, particularly the US, invading the countries concerned or engineering coups through support of local army units or other disaffected groups in the name of protecting the people from communism, killing the leaders and principal supporters of such reform movements and establishing ruthless police state dictatorships.

As long as such dictators protected neocolonial interests in raw material exports, they were allowed to benefit from continued state control of productive resources and tariff protections. They were encouraged to utilise aid, loans and raw material export revenues to increase their own military repressive power over their own populations and their own luxury consumption, rather than to develop infrastructure and industry that could ultimately compete with the West. It is in this context that we must understand the interventions of the International Monetary Fund (IMF) and the World Bank following the debt crisis of 1981. When these dictators were eventually removed or fled overseas, taking decades of aid and loan money with them, the burden of repayments fell upon the very people who had suffered as their victims.

The increasing neoliberal call for free trade has always been belied by continued protection and subsidisation of both the agriculture and industry of the developed world itself, reducing markets for developing world exports. Subsidised produce from the West was dumped in the developing world to the massive detriment of indigenous producers. Protection of infant industries in the developing world was increasingly undermined by threats and bribes from the IMF and the World Bank. Calls for the removal of 'tariff barriers' were actually calls to dismantle the tax base of countries relying upon such tariffs to fund infrastructure and social spending.

While ruthlessly repressing popular struggles for increasing democracy, human rights and equality, some developing world dictators sought to stabilise and enhance their own power through accelerating import substituting industrialisation (as in South Korea) and even some significant land reform and protections for the peasantry (as in the Philippines).

In the era of neoliberalism, and particularly following the fall of the Soviet Union, there was increasing hostility on the part of Western 'free traders' towards any such independent action, with a big push to replace remaining dictatorships with weaker, 'democratic' political forms, including periodic elections, allowing for more effective external control of government. Such regimes have typically lacked the strength or resolve to challenge odious debt, or resist takeover of their economies by foreign multinationals, the IMF and the World Bank, maintaining their populations in long-term debt and drudgery.

International trade – theory

As noted in Chapter 2, international trade law, as developed by the World Trade Organization (WTO), centres upon ideas and practices of the removal of 'restrictions' to free international trade, justified by the theory of comparative advantage. So do the WTO rules require moves towards complete freedom of investment, or free flow of capital around the world, without what is seen as government 'interference' or preferential treatment for nationally based businesses, and recognition and enforcement of 20 year US patents, supposedly in order to encourage and sustain innovation.

Poorer countries have been told that their comparative advantage lies in raw materials – subject to big supply and price fluctuations as a result of natural disasters and periodic overproduction, and substitutable by other products if prices do increase. While they pay the monopoly prices of the developed world cartels for manufactured imports, they have been prevented from forming protective cartels to ration output of such raw materials and have instead been forced into cutthroat competition with one another for desperately needed foreign currency to sustain imports and service debts.

The original theoretical model of comparative advantage involved capital mobility within a nation – to allow investment to shift into goods of comparative advantage – maintaining jobs and wages in the territory concerned, but not mobility between nations, allowing capital to move wherever costs can be most effectively externalised. Now, what is called 'world trade', is mostly planned exchange within transnational corporate structures, with 'prices' determined as much by tax minimisation as by any real-cost considerations.

Here, indeed, big corporations have taken advantage of globalised markets to minimise their taxation payments in a variety of different ways, thereby undermining government authority and provision of basic services. Transfer pricing enables them to claim that profits are generated in low tax areas, including offshore 'tax havens'. They also pressure elected governments, particularly the weaker governments of developing countries, to provide fiscal concessions in the form of tax breaks and subsidies.

The reality of comparative advantage

Comparative advantage certainly makes some sense as a possibility in some future, rationally planned world economy where benefits of specialisation are properly shared by all. However, the current reality is one of radical inequality of power.

Ricardo considered the unique supply environments associated with availability of mineral deposits and climate differences affecting agricultural production which can, indeed, necessitate international specialisation. But in mass production industries, now using the same technologies worldwide, as Keynes (Moggridge, 1982; Davidson, 2009, p. 131) argued, without enlightened planning on a world scale, the economic costs of self-sufficiency have to be balanced against the potential gains of 'bringing the producer and the consumer within the ambit of the same national economic and financial organization' [to assure full employment].

As Paul Davidson (2009, pp. 128–37) points out, in a world where multinational corporations easily and cheaply transfer technology and output across national boundaries, where some countries have outlawed sweatshop production while others have not, corporate leaders driven by shareholders or by greed to maximise profits without restriction, will increasingly transfer production techniques to the sweatshops, where real unit labour costs are significantly lower.

Such transfer is justified on the grounds of its driving the emergence of new 'higher value' jobs in the old, labour protected industrial heartlands, sustained by cheaper imports. This presupposes some limit upon the expansion of higher skilled jobs in low wage areas and/or some significant increase of world monetarily effective demand in line with the expansion in output made possible through involving such lower paid workers. Available empirical evidence provides little support for either such consideration.

Trade protection and high transport costs in the early postwar decades, coupled with relatively high labour costs in the developed world, encouraged ongoing technological innovation to reduce labour and resource costs per unit of output. The ready availability of cheap labour and cheap raw materials in the developing world in the neoliberal era has reduced investment in research and technological improvement, able to create new, high value jobs and to raise living standards across the board. This has left the West with employment increasingly reduced to 'industries that produce goods and services that are not tradable across national boundaries' (Davidson, 2009, p. 136). These jobs are weakly unionised and are often smaller-scale business operations, where workers are in a poor position to fight for better conditions. Intensified competition for such 'service industry' jobs amongst those displaced from the manufacturing sector by 'free-trade' has further helped to force wages down (Davidson, 2009, p. 136).

In the developing world, the increasing transfer of land from subsistence farming to more 'efficient' agricultural exporters, miners and electricity producers, and the destruction of import substituting local industries, produced by the removal of agricultural and industrial protection in the name of free trade, has created a huge pool of desperate, destitute unemployed people, which, along with repressive government control, has kept wages down to an absolute minimum.

Low wages in both developed and developing worlds have reduced monetarily effective demand, thereby restricting further investment and further job creation and continuing to further depress wages.

These developments are preventing efficient use of resources, with cheap labour obstructing resources and labour-saving innovation, and subsidising transportation technologies that pollute the environment and accelerate global warming. This is exactly the opposite of what David Ricardo had in mind when he first formulated the theory of comparative advantage.

Patent protection

As Nobel Prize winner Joseph Stiglitz points out, long-term patent protection impedes research based progress, by reducing the incentive to innovate. This is particularly the case where the system allows for minor variations on existing products to count as 'new' products and where it denies researchers access to knowledge crucial for further progress.

This is particularly evident in the pharmaceutical industry, the major driver of extended patent protection, with endless 'me too' drugs and very little genuine innovation. Real innovation has typically derived from public funding and research, with the private sector cashing in at a relatively late stage. Following acceptance of US patent laws in the developing world, life-saving drugs have increased two to three times in cost in the developing world. These issues are explored in later chapters, particularly Chapter 9.

Exchange rates

Floating currencies as another key element of the neoliberal deregulatory package, with values determined by international currency trading, are supposed to ensure balanced international trade. Countries in deficit find their currencies devalued, thereby increasing the cost of their imports and decreasing the cost of their exports, reducing the former and increasing the latter to eliminate trade deficits.

Whether this actually happens in practice depends, amongst other things, upon the price elasticities of demand for the exports and imports of the country in question. With relatively inelastic demand for key exports, a country can find itself with reduced export income (measured in terms of US dollars or euros) despite an increased volume of exports, because the increase has not been sufficient to offset the loss of earnings through the reduced sale price of such exports. With inelastic demand for key imports, again, a country could find itself paying much more (in US dollar terms) for less, because it is unable to reduce its import volume sufficiently to compensate for the increased costs. Devaluation can have disastrous consequences, actually creating a 'greater deficit in the export–import balance when measured in terms of a common currency such as dollars than that which existed before devaluation' (Davidson, 2009, p. 139).

As Davidson says, even with greater elasticities, a 'huge depreciation' of a deficit country's currency may be needed to reduce an unfavourable trade balance. And 'any large devaluation... will have a significant deleterious effect on the real income of the residents of the nation at least partly through creating inflation in the prices of all imports' (Davidson, 2009, p. 139). Such reduced demand will be passed on to exporter nations, increasing unemployment and lessening demand in the world market.

In practice, currency convertibility can expose countries to sudden and disastrous depreciation pressures. As Chang and Grabel (2005, p. 169) argue, the costs of non-convertibility, including the creation of black markets or bribery to acquire foreign exchange licences, are comparatively minor when compared to the threat of major financial crisis. Restricting convertibility can also control capital flight (through preventing owners liquidating their assets and taking the proceeds out of the country), discourage foreign ownership of domestic assets as a precursor to such flight, and reduce the ability of wealthy citizens to move their wealth abroad, rather than using it constructively within their own countries. Restricted convertibility gives governments the power to allocate scarce foreign exchange to priority sectors of the economy (Chang and Grabel, 2005, p. 169).

Western Europe and Japan adopted only limited currency convertibility in the 1950s and 1960s after substantial economic reconstruction, maintaining restrictions before this to avoid devaluation and capital flight. But the Western powers, acting through the World Bank and the IMF, have forced developing countries to adopt unrestricted convertibility much earlier in their economic development. Those countries, which have achieved substantial development later than that of Europe and Japan, particularly China, India and Taiwan, did so in part through maintaining restrictions on currency convertibility (Chang and Grabel, 2005, p. 170).

Deregulation has driven massive speculation in rapidly expanding international currency markets, and allowed rapid sale of particular national currencies in response to panics and prejudices, with sudden depreciations leading to cycles of further depreciation, declining asset values and financial crisis in the developing world.

Following the developing world debt crisis of the 1980s, the IMF's and the World Bank's structural adjustment program typically required debtor nations to devalue their currencies, and run independent central banks to supply overseas purchasers with as much of such devalued currency as they might want. Supposedly, as noted above, such currency devaluation increases import costs and decreases export costs, thereby eliminating trade deficits and increasing inputs of foreign currency to service debt and finance development. In particular, this was supposed to encourage 'technology transfer' through 'greenfield' investment.

In practice, along with demands for privatisation of all government operations, this has merely made all valuable national resources of these poorer, weaker countries available at knock-down prices to foreigners. Such resources are pumped out of these countries along with the profits from their exploitation, leaving devastation behind. Meanwhile, the costs of vital imports – including food and medical supplies – have escalated, with no government services or subsidies left to protect the poor majority. The legacy of free trade and investment is a billion people without adequate food, water or housing, let alone education or medical care, with tens of millions more joining them each year. Debts remain unpaid with interest mounting up.

China

As noted earlier, a few developing countries have been able to pull themselves out of rural poverty following the Second World War. Those that have done so have achieved this, not by embracing free trade and investment, but by recognising the need for a strong central state control of surplus and industrial development built upon rapidly achieving significant scale economies in key industrial sectors. They have also had the advantage of significant US financial and strategic support. Through strong central control of very substantial agricultural resources, China has been able to follow a similar path without benefit of external financial assistance.

China has undergone a process of rapid and large-scale export led industrialisation, selling goods to the developed world, particularly the US, by utilising the surplus from the agricultural revolution to sustain the expansion of an urban industrial work force, keeping labour costs low. Only when the process was well advanced did the Chinese leadership proclaim their commitment to free trade and investment and join the WTO, presumably to try to avoid US import restrictions and demands for significant upward valuation of the Chinese currency.

China has been the only country allowed to maintain the non-convertibility of its currency – the yuan – with its value set by policy makers, to gain optimum advantage in relation to other currencies, rather than being set by market forces. An undervalued yuan has contributed to exports of jobs from west to east and to China's trade advantages. China has notoriously failed to protect copyrights and patents on overseas products. The central government has given subsidies, tax rebates and credit support to export industries and subsidised locally produced goods through a coupon system. It has restricted exports of raw materials where it is a major world supplier.

Along with Japan and some other East Asian countries, China has been strong enough to try to protect itself from fluctuating currency prices by accumulating ever greater reserves of what have been seen as safe currencies – mainly US dollars. The Chinese central bank has accumulated hundreds of billions of US dollars to prop up the value of the dollar relative to its own currency, and thereby maintain its competitive advantage in trade with the US.

China's reserves mainly take the form of short-term US Treasury bills, which can be sold quickly for cash but have very low interest rates. This means that the US is, in effect, borrowing a lot of money from China at very low cost, to buy Chinese manufactures. More jobs in export industries in China have meant fewer jobs in the US and other parts of the developed world. In addition to the dire social consequences for developed world workers, this has created problems for developed world businesses trying to sell goods and services to such workers. This, in turn, has contributed to attempts to boost demand at home through easy lending and low interest rates – contributing to asset inflation – particularly in the US housing market.

Low interest rates for US bonds have meant that not only China and Japan but also poorer East Asian economies holding such bonds have been losing out on the higher interest rates (or profits) that could have been produced by other uses for such funds, while the US has twice funded 'huge tax cuts' – mainly for the already rich – out of such cheap loans. The rest of the world has suffered as a consequence of the huge loss of aggregate demand produced by 'removing US$750 billion of purchasing power from the global economy every year' – locked up in reserves (Stiglitz, 2006, p. 251). This has exacerbated trade imbalances, with such funds unavailable for the purchase of goods of deficit nations. It has allowed the US to run up huge trade deficits, in effect, funded simply by printing money, while poorer deficit nations have been persecuted through structural adjustment.

The bigger the US deficit has become, the more this has undermined confidence in the US dollar as reserve currency. 'Already by 2005 . . . China had moved about a quarter of its reserves out of dollars' (Stiglitz, 2006, p. 254). A rapid sell-off of dollars would quickly depress the value of the dollar, thereby devaluing the as yet unsold dollars and undermining Chinese exports to the US. Indeed, private capital flows to the US collapsed over the summer of 2007 with the dollar's value falling rapidly after this. 'By early 2008 import prices were rising at an annual rate of 13.6%, the fastest increase for more than 2 decades' (Turner, 2008, p. 92). Not all Europeans are happy about an increasing move into the euro. As Stiglitz (2006, p. 256) points out, 'as central banks hold more euros as reserves, the value of the euro will increase, making it harder for Europe to export and opening it up to a flood of imports'.

The increasing shift of manufacturing industry from west to east, to take advantage of cheap labour costs enforced by an authoritarian Chinese leadership, has maintained a powerful downward pressure on jobs and wages in the West, particularly the US. This has reduced pressure towards increased productivity, through innovation, and has increased pressures towards debt, rather than wage, led consumption. It has meant big profits for producers at home and abroad, through increased exploitation of the labour force. However, opportunities for genuinely productive investment have been reduced by limited monetarily effective demand. Surplus profits have increasingly moved into speculation in financial markets. These factors have played a central role in producing the current crisis.

Social consequences

In the developed world, first increased unemployment and underemployment, created by interest rate rises, privatisation and legistative restriction of trade union activity in the 1980s and 90s, then increasing export of manufacturing industries to selected regions of the developing world, radically reduced workers' power and living conditions. In the English-speaking world, in particular, there is a widespread fear of unemployment, driven by reduced social welfare provision and increasingly effective internalisation of neoliberal ideology. This is vigorously pushed by the privately controlled mass media and has further weakened workers' capacity to fight for better conditions, leaving them prey to longer hours of work for less, or increasing casualisation.

On the other side of the class divide, the reduction of workers' power, both in the developed and the developing world, has increased the power and wealth of those controlling workers' pay and conditions and those living from profits rather than wages. The gap between the richest and poorest 10% in the developed English-speaking world has increased by 40% since 1975 to reach a level 'unprecedented since records began' (Wilkinson and Pickett, 2009, p. 234). In the US, the richest 20% have increased their income to the point where it is more than eight times greater than that of the poorest 20%. In the UK and Australia, the income of the richest 20% is around seven times greater.

There is now substantial evidence linking inequality in the developed world with major physical and mental health problems, with increasing inequality leading to increased levels of avoidable mortality and morbidity. Inequality is strongly linked to such social problems as drug and alcohol addiction, homicide, lack of trust, obesity, imprisonment, adolescent behavioural problems, including involvement in crime, and reduced social mobility. The causal connection between developed world inequality and adverse health outcomes is explored in detail in Chapter 4.

In the developing world, the effects of increasing inequality in the neoliberal period are complicated by the consequences of increasing absolute poverty that leaves hundreds of millions of people in destitution, with tens of millions more joining them each year. The health consequences of such absolute poverty are readily apparent. These issues are explored in Chapters 4 and 5, including consideration of the underlying causes and consequences of such absolute poverty.

Disadvantages of financial deregulation

Neoliberals claim that all people benefit from financial deregulation – from the process of allowing allocation of financial resources through the operation of market forces – without government intervention in the financial sector. As noted in Chapter 2, it includes deregulation of interest rates, removal of credit controls, privatisation of government banks, free entry of foreign banks and financial institutions into domestic financial markets, and market based instruments of monetary control.

In fact, financial deregulation has really only benefited selected groups at the expense of the great majority. Some larger firms 'have received significant finance through capital markets created or expanded by financial deregulation, cheaper than that available via bank loans', but this has increased their market power at the expense of smaller firms denied such access, and has fuelled rampant speculation. The expansion of global 'financial integration increases systemic risk, financial fragility and volatility and increases the potential for financial crises' (Chang and Grabel, 2005, p. 153).

Far from intensifying competition and competitive price reduction by banks, such deregulation has facilitated increasing centralisation and concentration through international takeovers and mergers, as well as cartelisation, allowing monopoly price-fixing of bank charges.

This development is not confined to the banking industry. Opening up a national economy to unrestricted foreign investment (particularly in a context of free currency markets which can rapidly devalue a nation's currency) allows all of its businesses (and resources) to be taken over, at potentially very low costs, by massive transnational corporate operators, leading to job loss through rationalisation, repatriation of profits and monopoly pricing of output. Economies weakened by currency devaluations are unlikely to be in a strong position to impose heavy taxes upon such foreign investors or requirements that they pay living wages or operate sustainable production processes.

For the less developed world, complete freedom of foreign investment just means that whatever surplus can be forced out of the working population and the natural geography goes into the pockets of foreign investors, rather than being channelled into the future development of the country through building up health, education, infrastructure, and technological advance.

Empirical evidence shows that financial liberalisation around the world has failed to increase domestic savings, it 'has not promoted long term investment in the projects and sectors central to economic development' (Chang and Grabel, 2005, p. 154) or a reduction of social problems, including unemployment. As Chang and Grabel point out, it has created 'the climate, opportunity and incentives for investment in speculative activities and a focus on short term financial as opposed to long term developmental returns'.

Only at the time of original issue do stocks or bonds actually raise investment capital or pay off debt for companies. Secondary buying and selling has no direct impact on the company that issued them. Many share purchases are about speculation – attempting to buy low and sell high – rather than long-term support of particular corporations. At the same time, continued dividend payments represent an ongoing drain on funds that could be used for new investment and jobs, better wages and conditions, and research into new, safe and sustainable technologies.

Efficient share markets – reality

Keynes clearly identified the obvious problem with the theory of efficient share markets underlying neoliberal financial policies. It assumes, as Steve Keen (2001, p. 234) says, that there is no feedback from share market valuations to investor perceptions. It is certainly true that insider trading, based upon such relevant 'inside' information, is widespread and allows those in positions of power, already enjoying salaries and bonuses many times greater than average wages, to further benefit at the expense of 'mum and dad investors'. The regulators notoriously fail to address such injustice.

But apart from such short-term inside information, there is very little other data on which to make any medium- to long-term judgments of company performance. Investors extrapolate current trends into the future. Professional fund managers look at one another to try to predict 'how the majority will value particular companies in the immediate future'. Individuals respond to the fads and fashions which hold others in thrall, as to which particular 'indicators' are 'fundamental' and which are not (e.g. national balance of payments figures).

'A rising market will tend to encourage investors to believe that the market will continue rising' (Keen, 2001, p. 239). Even if they don't believe this, there will still be pressures to buy, to cash in on such rising prices. 'Such a market can find itself a long way from [anything that could be called] equilibrium' as 'self-reinforcing waves of sentiment sweep through investors'. Of course, eventually, and inevitably, such waves break, when it becomes clear that 'valuations have gone far beyond what is sustainable by corporate earnings'. Insofar as regulators strive to maintain 'confidence' and promise to bail out failing banks, they can encourage excessive risk taking and contribute towards such uncontrolled expansion and subsequent collapse.

Share markets interact in complex ways with the real economy, responding and contributing to cyclical fluctuations, interest rate changes and regulatory interventions. Money flows into increasingly speculative channels when inadequate demand reduces the profit potential of real investment.

Minsky

In the 1970s, Keynesian economist Hyman Minsky provided a realistic and relevant analysis of the interaction of business and finance to produce cycles of increasing instability. Minsky describes an upturn, started by some positive structural development, with increased investment and employment, involving rapid growth by firms whose cash receipts generally cover their cash outlays. At this stage there are low levels of lending by risk averse banks, who remember past financial failures. However, such successes encourage others to become involved, who see increased borrowing – or 'leverage' – as a means to improve upon the yields earned by the 'cash players'. Managers and bankers come to see existing debts as unproblematic and rewarded by growth. A general decline in risk aversion by investors and bankers sees more optimistic evaluations of investment prospects and increased availability of credit and 'an accelerated growth of borrowing'. As Keen (2001, p. 252) says, with regulation, this leads to more non-bank financial intermediaries coming into being, otherwise, there is a big increase in money created by the banks.

In a 'euphoric economy' 'asset prices are revalued upwards', low yielding financial instruments are devalued, leading to increased interest rates. Borrowing continues to accelerate as a result of continued inflation of asset prices – with high returns still expected from speculative investments. Such euphoric conditions allow for the proliferation of what Minsky calls 'Ponzi firms', trading assets in a rising market, and incurring significant debts in the process

> The servicing costs for Ponzi debtors exceed the cash flows of the businesses they own, but the capital appreciation they anticipate far exceeds the interest bill. (Keen, 2001, p. 252)

Their borrowing and spending further push up interest rates and asset prices which increasingly undermine other businesses, forcing them to sell assets. This ultimately reverses the asset price boom, and leaves the Ponzi financiers with assets that they can no longer trade at a profit. The banks that lent to them find their major borrowers defaulting and try to compensate with further interest rate rises. Asset prices go into free-fall and the boom turns into a slump. The applicability of this analysis to events thirty years on from the time of its formulation is apparent.

History

The stock market crash of 1929 and subsequent bank collapses and recession, were fuelled by interest rate cuts that provided cheap money for investment banks, the growth of investment trusts and massive leverage, with ordinary deposit taking banks increasingly drawn into the speculation. The Roosevelt presidency of 1933 issued in a new period of increasing regulation.

In particular, as Mason explains

> The Glass-Steagall Act of 1933 made it illegal for a bank that holds people's savings to engage in speculative activity. It also introduced the first federal scheme to insure bank deposits, the Federal Deposit Insurance Scheme. As a result the investment banks' access to capital was reduced. Meanwhile laws preventing the consolidation of banks across different states ensured that the banks that did have access to ordinary people's savings could not become huge. (2009, p. 61)

This meant that in the postwar boom, investment banks, as rich people's gambling clubs, remained clearly separated from ordinary deposit taking banks. Capital and exchange controls restricted and directed investment, and the flow of money. Such intervention and regulation contributed to the boom through keeping consumption – including a substantial social wage – in line with productivity increases produced by new technology.

The disaster of the 1930s, with mass unemployment and financial chaos driving the rise of fascism, and thereby paving the way for the Second World War and the Holocaust, led even right-wing politicians to accept the necessity for deep-going government market intervention and regulation in the period after 1945. Unlike earlier state intervention, the goal was now seen as averting the worst consequences of cyclical fluctuations of the economy and promoting full employment.

A new period of crisis in the early 1970s provided the opportunity for the neoliberal reformers, hostile to such intervention and regulation, to begin to dismantle it, weakening workers' organisation through engineered unemployment and legal restriction of trade union power, winding back the welfare state, and pushing forward privatisation and financial deregulation. But when, by the 1980s, most Western governments had abandoned the pretence that full employment was still a desirable social goal, 'tax breaks and subsidies for the private corporate sector became bigger and more widespread than ever' (Shutt, 2001, p. 56). As the increasingly pro-business media and political establishment derided corporatism, welfare and government spending in general, more and more of poorer taxpayers' money flowed into corporate welfare, propping up the unearned dividend income of the wealthy and the inflated remuneration of business executives.

Commercial banks were allowed to merge and consolidate. Between 1987 and 1996, the Glass–Steagall rules were dismantled. The final repeal of the *Glass–Steagell Act* in 1999 left retail banks free to underwrite and sell securities. And in 2004, the US Securities and Exchange Commission permitted banks to increase their leverage ratios – of liabilities to net worth – from ten to one to thirty to one.

At the same time, laws preventing the merger of banks in different states were repealed, as was the insurance/banking split. The banks then restructured themselves to get around the remainder of the existing rules. (Mason, 2009, p. 62)

Emerging mega-banks became increasingly involved in worldwide speculation in currencies and derivatives, with the value of the financial economy ultimately overtaking and massively outdistancing the value of the real world production upon which it was supposed to be based.

Meanwhile, central banks continued to provide 'safety nets' for the financial sector as lenders of last resort, as originally established to protect against a 1930s style banking meltdown. In the context of deep-going neoliberal financial deregulation, this created an ever greater temptation for financial and non-financial enterprises to make risky speculative investments in the expectation that they would make a fortune, if successful and would lose 'little more than their jobs if the gamble failed' (Shutt, 2001, p. 56).

Derivatives

Chapter 2 touched upon the way in which orthodox theory views all risk as being statistically predictable in character (due to earthquakes, bad weather etc.). It sees the continuing development of new financial instruments and markets as the best way for the economy to cope with risk by allowing investors to 'hedge' against losses in increasingly sophisticated ways.

We saw how orthodox theory compares futures contracts with insurance – spreading risk and loss over larger groups of people. Other such derivatives include the sale of options (to buy), rather than of actual things, and agreements to swap commodities, shares, bonds, contracts or currencies and their earnings for a set time period.

But it is typically speculators, including hedge funds, private equity funds and investment banks – as the gambling clubs of the ultra-rich – who are engaged in large-scale derivative trading. Futures contracts can be very highly leveraged, meaning that speculators can resort to substantial debt relative to their equity in capital structure to finance the purchase of derivatives, creating big possibilities for gain but also big possibilities for greater loss. As Peter Wahl (Kohonen and Mestrum, 2009, p. 75) points out, 'the Carlyle Capital Fund, which collapsed in 2008, conducted transactions for US$22 billion with only US$670 million of its own assets; this is a leverage of around 1:32'. McKinsey estimates that the leverage of the 9000 or more hedge funds operating in 2007 was at least three to four times their assets under management of US$1.7 trillion (Kohonen and Mestrum, 2009, p. 75).

Low interest rates encourage such leverage. Far from neutralising risk, derivatives create new systemic risks due to the complexity of operations and structures involved. These derivatives undermine the capacities of both speculators and regulators to monitor and manage risk. Due to an increasing risk aversion rising with an increasing scale of borrowing and indebtedness, this leads to reduced liquidity.

The proliferation of instruments increasingly far removed from the reality of current production – as advocated and encouraged by neoliberals – renders real values increasingly uncertain and incalculable. Insurers had no real basis for assessing the risks of these new instruments but offered comprehensive insurance nonetheless. In 2002, Warren Buffet (Mason, 2009, p. 66) said 'central banks and governments have so far found no effective way to control or even monitor the risks posed by these contracts. In my view derivatives are financial weapons of mass destruction . . . '.

According to Paul Mason (2009, p. 66), the global derivatives market stood at $370 trillion in June 2006. By December 2007, it had reached $596 trillion. Foreign exchange futures doubled in this period and credit default swaps (CFS) – whereby banks and hedge funds insured each other's loan and security portfolios – grew from under $1 trillion in 2000 to $58 trillion in 2007. The total value of companies in the world's stock markets in 2007 was $63 trillion – close to the world gross domestic product (GDP); the value of derivatives was $596 trillion. Total currency traded was $1168 trillion, which was 17 times the world's GDP.

Easy money

It was earlier noted that, while neoliberal regimes have pushed up interest rates to forestall any possible improvements for labour, they have also reduced them in face of any serious threat to capital. Interest rates have been sharply cut to pump liquidity back into the market to respond to threats of market meltdown produced by neoliberal deregulation.

After the 1987 stock market crash, the second Gulf War, the 1994 Mexican crisis, the 1997–98 Asian financial crisis, the collapse of the Long Term Capital Management (LTCM) hedge fund in 1998 and the 2000–01 bursting of the dotcom bubble, federal fund rates were significantly lowered. As Kevin Rudd says

> investors increasingly came to believe that when things went bad, they would be protected by monetary policy – through low interest rates, high liquidity and the protection of asset prices . . . [This] added yet more fuel to the fire, in the form of cheap money available for lending. (2009)

In the business world, the easy lending fostered by Western governments, as a response to increasing financial market volatility has fuelled mergers, takeovers and acquisitions by private equity funds. This has further concentrated corporate power and accelerated the overheating of the markets, as demand has pushed up asset prices. It has encouraged 'leveraged buy-outs'. This is where a significant percentage of the purchase price of a company is financed through borrowing. The assets of the acquired company are used as collateral for the borrowed capital, followed by asset stripping and 'rationalisation', including sackings, to pay off such debts (see Morris, 2009, pp. 137–8).

In the consumer goods market, mortgage lenders in the US, with access to free money, relaxed their standards and invented new ways to stimulate business and generate fees. With incomes stagnating in the new millennium, particularly in the US, as a result, in part, of capital export to China, lenders used clever tricks to try to make houses appear more affordable to increasing numbers of people. The most popular devices were adjustable rate mortgages with below market rates for initial two-year periods. As George Soros (2008, p. xvi) observes, 'credit standards collapsed and mortgages were widely available to people with low credit ratings – so-called sub-prime mortgages'. Many of these sub-prime loans were for the full price of homes, and went to people without assets or income.

While stock prices fell from 1997–2006, US house prices increased by over 120%. From 2000–05 there was a demand driven increase in the value of existing homes of more than 50% and an acceleration in home construction. With house prices rushing upwards, loan defaults were seen as unproblematic, since lenders would get more from foreclosure.

Risk reduction

In keeping with the orthodox theory of risk reduction through financial innovation, George Soros explains

> Banks sold off their riskiest mortgages by repackaging them into securities called collateralized debt obligations. CDOs channelled the cash flows from 1000s of mortgages into a series of tiered... bonds with risks and yields tuned to different investor tastes. The top-tier tranches... would have first call on all underlying cash flows, so they were sold with an AAA rating. The lower tiers absorbed first dollar risks but carried higher yields. In practice, the bankers and the ratings agencies grossly underestimated the risks inherent in subprime loans... (2008, p. xvii)

Loans were increasingly organised by individual 'brokers', rather than banks or building societies, on a fee-for-service basis, then 'warehoused' by 'thinly capitalised mortgage bankers', before being sold to investment banks (e.g.

Lehman Bros, Goldman Sachs) that created the collateralised debt obligations (CDO) for sale to institutional investors (e.g. pension funds, overseas banks), supported by credit rating agencies.

As Soros notes

> Starting around 2005, securitisation became a mania... Investment bankers sliced up CDOs and repackaged them into CDOs of CDOs. The highest slices of lower rated CDOs obtained AAA ratings.... Securitisation spread from mortgages to other forms of credit. (2008, pp. xvii–xviii)

Federal Reserve inflation-fighting interest rate increases of 1% to 5.25% from 2004 to 2006 were the straw that broke the camel's back, leading to massive falls in house prices. With these various deals generating increasing tens of trillions of dollars upon a foundation of unpayable debt, things began to fall apart first in the US mortgage lending businesses, registering multibillion dollar losses with increasing numbers of mortgage defaults. This was followed by mortgage hedge fund collapses. Investors – led by big hedge funds – rapidly shifted from property (as the housing market crash accelerated) and into commodities, including oil, metals and foodstuffs, leading to massive new price inflation, with futures contracts in these areas jumping from $2 trillion in 2004 to $9 trillion in 2007 (Mason, 2009, p. 107). Along with other developments, considered in Chapter 5, this led to massive food price increases, plunging millions more poor people around the world into desperate food insecurity. Many banks responded to the commodity bubble by pushing up interest rates, creating further problems for debtors.

In August 2007, financial markets moved into panic mode. Investment banks with large amounts of CDOs – and large loan commitments to finance leveraged buy-outs – lost tens of billions of dollars. This, in turn, unnerved the stock market and price fluctuations became 'chaotic' (Mason, 2009, p. 107).

All this put huge pressure on the banking system. With 'good' debts bundled together with 'bad', banks had difficulty assessing their exposure and even greater difficulty estimating each other's exposure. Consequently they were reluctant to lend to each other and eager to hoard their liquidity (Mason, 2009, Chapter 6).

After August 2007, central banks, that had been maintaining high interest rates to fight inflation, started to pump hundreds of billions of dollars into the banking system, while they kept interests rates high. The US stood alone in slashing interest rates to 2% during the credit squeeze while also delivering tax cuts to try to stimulate the economy (Mason, 2009, p. 112).

Billions of central bank dollars contributed to a writing-off of huge amounts of bank debt. But central banks found it difficult to pump enough liquidity into the system to keep it functioning. Such actions failed to stop a worldwide credit squeeze. The British mortgage bank Northern Rock collapsed, leading to a government bailout (ultimately nationalisation at a cost of £100 billion) and a guarantee of the safety of accounts in other British banks (Mason, 2009, pp. 103–4).

Policy

With the banks generally failing to raise more capital for recapitalisation, and failing to start lending to one another again, by early 2008 stock markets were plummeting. By July 2008 the UK economy was in recession, and in autumn 2008 recession had hit the US economy. On 7 September Fannie Mae and Freddie Mac, major government backed mortgage lenders, were taken over by the US Government to prevent collapse, with managers sacked and shareholders wiped out.

As Mason says

> by the autumn of 2008, the futures market for commodities was reflecting the expected onset of recession. But it is also clear that billions of dollars were pulled out of the commodities markets as the meltdown of September–October forced both hedge funds and large institutions to 'de-leverage', that is, to call in and pay back the loans that had been fuelling one speculative bubble after another. (2009, p. 109)

In October there were coordinated interest rate cuts by central banks around the world indicating a reassertion of significant state control, but substituting more cheap money for active state regulation of finance. The US reduced rates to zero in December, and started fabricating new money to purchase debt, a tactic which had contributed to the stagnation of the Japanese economy since 2002.

Following earlier multi-billion dollar 'rescue packages', including US$170 billion in bailout funds to banks and insurance companies (US$165 million of it handed out in bonuses to executives of the AIG group), the Obama Government's key response to the deepening crisis was a plan for taxpayer subsidised buyout of the banks' 'toxic assets' by hedge funds, private equity groups and insurance companies. Low cost loans and subsidies were presented by the Treasury and the Federal Reserve to hedge funds and private equity groups to fund the purchase of the banks' 'junk assets' at massively inflated prices (to the tune of US$500 billion to US$1 trillion) (Grey, 2009).

The idea is that bank stocks rise with toxic waste cleared off the books, while private investors either make a big profit selling the assets when the housing market recovers, or get bailed out by taxpayers (with loans guaranteed by the government). As Grey argues, rather than facing public attack for buying such assets themselves at hugely inflated prices, the government is allowing the 'market' to set such inflated prices through its subsidisation of private investors.

The Federal Reserve's huge money creation scheme – with 'the buy out of $1.15 trillion in securities, including $300 billion in long-term government debt' to 'prop up the value of US Treasury notes', threatens to significantly reduce the value of the US dollar and thus of the US$1.3 trillion dollars of reserves still held by China. It is not surprising in this context that Zhou Xiaochuan, head of the Chinese central bank, should call for the replacement of the US dollar as the global reserve currency in March 2009 (see http://www.wsws.org/articles/2009/mar2009/chin-m24.shtml).

Even staunchly neoliberal regimes have responded to the threat of an accelerating multiplier of reduced employment and investment, completely undermining their national economies, with substantial deficit spending by state authorities to maintain employment. In Australia, the supposedly more left-wing Rudd Labor Government is spending tens of billions of Australian dollars in such 'stimulation'. There are crucial issues of here of how this spending has been directed, which are addressed in later chapters, particularly Chapter 7. Only South Korea has targeted the bulk of stimulus spending to new low carbon technologies and climate change protection. The Australian effort has radically failed to do this.

A major danger is that an accelerating cycle of falling asset prices can push up the real debt burden and increase defaults. Financial institutions in trouble, due to inadequate borrowing, can still raise real borrowing costs to try to compensate for inadequate demand even as central banks reduce key lending rates. Reduced credit availability or increased cost of credit can still further reduce real demand, pushing asset prices still lower, causing more business collapses, more unemployment and still further reduced demand. Falling asset prices can scare off buyers, fearing rapid devaluation of their investment. With falling interest rates and asset prices, potential buyers can hold off, expecting further falls.

Ethics – problems of just deserts

From a utilitarian perspective, the factual problems already touched upon provide a comprehensive moral indictment of neoliberal theory and practice. Neoliberal policies radically fail to maximise social welfare (or happiness) on a world scale. Compared to other possible policies, which could have been – and could still be – adopted, they radically fail the utilitarian test of welfare (or happiness) maximisation.

The situation is no better in respect of issues of distributive justice and fairness. The prioritisation of property rights per se, by Locke and Nozick, is simply a failure to properly address the issues. It provides no real justification for the radical inequality created by unregulated competition and no justification for identifying 'liberty' with private property protection alone. There is no solid evidence provided to support claims of inevitable totalitarian interference as a necessary condition of greater equality.

Looking back for some sort of legitimate 'original appropriation', we typically find colonial and neo-colonial theft and murder on a grand scale. Subsequent market transactions are saturated with coercion and misrepresentation of various kinds. The wage contract, in particular, can hardly be said to be freely entered into by individuals with no property and no other options for access to basic necessities of life.

Chapter 2 considered neoliberal support for the idea of just deserts – and the idea that free capitalist markets and corporations – governed and regulated by various legal systems – basically deliver such just deserts – with individuals rewarded in proportion to their valuable productive contribution as measured by the market price of the goods they produce. In this chapter, we have seen good reasons as to why we should not equate market prices with real values to customers or real social costs to the community.

In practice, market prices are determined on the supply side by monopoly power, and cost externalisation on the demand side; by psychological manipulation; and very unequal income distribution, with radically restricted 'choice' on the part of the majority of purchasers. In the absence of legally linking wages to productivity, actual rewards for unskilled labour depend on the bargaining power of labour. This in turn depends on levels of unemployment, underemployment and the legal powers of trade unions.

Rewards for skilled labour are closely linked to issues of access to the relevant skills. These are restricted by cost and organised limitation of available training rather than the real productive contributions of labour.

In the developing world, in particular, high levels of unemployment and legal restriction of workers' rights radically reduce workers' bargaining power. This allows employers to enforce low wages despite long hours and high levels of real material output by workers. In a situation where subsistence farming has been destroyed through cheap imports and through the expansion of export agriculture, mineral extraction and electricity generation, this has created vast pools of desperate unemployed people. These people do not have even basic social welfare provision, and the attempt to fight for better wages and conditions is frequently met with brutal repression. It is very much a buyers' market for labour, with thousands of desperate people competing for each job, keeping wages at rock bottom. Goods produced for a few cents of labour costs sell for 10s or 100s of dollars in the markets of the developed world. However, they can still undercut the costs of goods produced by first world labour and so such developing world slave labour puts continuous pressure on pay and conditions in the developed world.

Productivity depends more on access to technology than on any particular contribution of human capital. Yet individual workers have no power to determine the quality or quantity of technology available to them. Again, their capacity to win pay increases, as a result of increased productivity, depends upon the state of the labour market and of the legal protection of workers' rights, including rights of collective action and collective bargaining.

Most productive activity in a developed industrial society is intrinsically collaborative, with a high level of division of labour within and between enterprises. This makes it impossible to tease out or quantify individual value creation within or between professions or class groups. CEOs claim responsibility for increased profit and market share, when in fact it is partly a function of the skills of the armies of specialists at their disposal, partly a function of their ruthless exploitation of workers, consumers and of the natural world, and partly a function of social and natural forces beyond their control. At the same time, they disclaim responsibility for failure – blaming such objective forces, and taking their inflated salaries, bonuses and golden handshakes anyway.

The orthodox neoliberal picture is on the right track by associating restricted supply of productive skills in demand as a mechanism maintaining high prices for the exercise of such skills. Such restricted supply is a result of access to the skills in question being restricted – directly or indirectly – by inherited wealth, social situation and power of professional associations and governments, and by the good fortune of being born in the developed world, rather than by any lack of innate ability, or unwillingness to struggle or sacrifice.

Rewarding labour as sacrifice

More socially minded liberals tend to be sympathetic to the idea of reward, in proportion to effort and sacrifice, rather than value creation. The ethical basis for such a preference is clear. Whereas it is unfair to reward or punish individuals for considerations beyond their control (like innate abilities or access to technology), it makes sense to reward their free decisions to devote time and effort to socially valuable activities, in proportion to the extent of effort and sacrifice involved. This might, indeed, be a defensible principle of remuneration in a future planned world economy.

We see problems with this idea when we try to relate it to the current social situation. Some people find their work intrinsically rewarding and therefore would not seem to require or justify extra reward, while for others it is pure drudgery. Yet the former are generally much better paid than the latter.

Those who can live comfortably without working can, indeed, be 'sacrificing' enjoyable leisure pursuits to engage in particular productive activity.

Those who cannot, could be said to be sacrificing only poverty and social marginalisation in order to go to work on the same job for a wage. This cannot mean they should therefore be paid less to do the same work as the wealthy and the privileged. Just because people are not involved in the paid labour market, producing commodities for sale, this by no means implies that they are not making a constructive social contribution and deserve no reward. With limited job opportunities, those who choose to refrain from participating – or are prevented from doing so – are providing work for those who want it. Continued consumption by the unemployed, on a basis of regular social security payments, maintains demand for the output of the employed and reduces the likelihood of a spiral of deepening recession.

Long-term unemployment undermines physical and mental health and individual capacities for productive effort and sacrifice in the future. If productive labour is itself a human need, and the unemployed are denied fulfilment of that need, then they deserve compensation rather than further deprivation. Perhaps the idea of reward for sacrifice could be developed to cover all of these considerations. It certainly needs to be pushed beyond a restricted conception of reward for labour.

Capital as sacrifice

The problems of the 'effort and sacrifice' idea are even more obvious in relation to provision of physical capital. Someone who already has all the material goods for a comfortable and protected life, but uses extra wealth to generate yet more wealth, is not making much of a sacrifice. Yet these are the people who get the really big rewards of capital appreciation and dividend income around the world. The 'capital as sacrifice' idea suggests that the more surplus wealth these people use in this way, the bigger the (added) reward they deserve. However, this makes no ethical sense at all.

As noted earlier, few purchases of securities actually provide any funds for expansion of production, since they involve purchases from other shareholders. As Marjorie Kelly (2003, p. 2) points out, 'among the Dow Jones industrials, only a handful have sold any new common stock in 30 years... in recent years about 1 in 100 dollars trading in public markets has been reaching corporations'. Share ownership drains resources from the productive economy, through far greater payouts of dividends than original funding of productive activities.

It might be the case that some people have made genuine sacrifices (worked hard in unrewarding but socially valuable jobs) in order to buy original offers of shares to fund useful productive investment. However, they still do not deserve rewards of dividends in perpetuity for no further contribution. This is taking value from the actual productive workers who generate it, and from future investment, to expand or humanise productive activity, and giving it to those who are no longer contributing anything.

This points to the real social function of share markets – as tools for channelling the wealth created by the workforce into the pockets of the rich, under the guise of 'investment funding'. While such markets contribute little to the actual funding of real productive activity, they can very effectively undermine such activity, with stock market collapses destroying real production and real wealth on a vast scale, working people being the principal victims.

On the face of it, these considerations appear to be even more significant in relation to income derived as interest on bank deposits – as recognised by the prohibition of such interest in Islamic law. We must not neglect the central role of the banking system in producing the current crisis. However, it is important to see that the crisis derives not so much from the use of bank deposits to provide funds for productive investment, as from the creation of new loans, as debt, by the banks themselves, and the use of such new money for speculation – in the Ponzi phase of stock price inflation – and wealth destruction – through leveraged buyouts and rationalisation, rather than for the creation of new wealth.

The virtually unregulated creation of new money as debt by the banking system has been a major contribution to the current crisis. And as a major driving force of unsustainable growth, it is something which must be seriously addressed in the future. Bank deposits, by the working population with interest payments, could provide a much more stable, rational and fair system for funding productive investment than stock markets sustained and driven into crisis by the excess wealth of the ruling class.

Underdevelopment and inheritance

Underlying most of the problems considered so far is the radical failure of equality of opportunity in contemporary market society. Social arrangements are in place to ensure that some people start out with significant wealth creating opportunities denied to others. Yet without genuine equality of opportunity, any ideas of reward for productive contribution make no sense at all.

Internationally, the radically varied levels of economic and technical development of different nations and regions create quite contrasting patterns of opportunity for people born in different parts of the world. The great majority of people born in a developing country have radically restricted opportunities compared to those in the developed world.

Nationally, unrestricted inheritance and other forms of undeserved property acquisition on the one hand, and the failure of unregulated capitalist markets to ensure full employment on the other, create dissimilar patterns of opportunity for people born into different social locations. Such inheritance of physical capital is completely at odds with the ideas of just deserts or sacrifice since nothing has necessarily been done by the recipients to earn such wealth. Through interest, capital appreciation, rents and dividend payments, inherited wealth generates further wealth without any necessary contribution on the part of the property owner.

By virtue of parental privilege, some people grow up in protected environments and lifestyles, with access to high quality health care, training, 'contacts' and 'networks' denied to others. Such privileged access gives some individuals special value creating powers denied to others. At the other end of the social scale, failure of the capitalist market system to generate full employment means that some people are not only denied the benefits of inherited wealth and privileged upbringing, but are also denied any opportunity to acquire the necessities of life or contribute constructively to the material life of the community.

This suggests that, in the present system, reward in proportion to the (market) value of goods produced or to the extent of effort or sacrifice involved in the production of goods for sale, whether by physical or human capital, whether or not they actually exist in practice, are radically unfair and unjust. They provide no kind of moral foundation for a defensible system of organisation of production and distribution.

Social mobility

Chapter 2 showed how some proponents of the current system argue that high levels of social mobility offset such initial inequality of opportunity. This mobility allows all who choose to take appropriate steps of skills acquisition and exercise to be able to radically increase their value creating capacities and hence rewards. Yet such proponents are typically virulently opposed to any sort of genuinely free mobility of labour across national boundaries to give workers from the developing world any chance of upward mobility through access to skill acquisition and job opportunities in the developed world.

In a capitalist system, any such accelerated flow of labour to the developed world would produce some upward pressure for better wages, conditions and opportunities in the developing world. Without concerted action by the working class, it would produce yet further downward pressure on wages – conditions and opportunities – in the developed world.

There is plenty of evidence that such mobility within the developed world itself is also largely illusory. Whenever rich lists are published, at least half of those listed have directly inherited their positions, and half of the rest have started with millions. The basically pyramidal class structure of contemporary society, with the number of available positions declining as rewards increase moving upwards through the hierarchy, means that no matter how much effort the majority might make, there is no way for more than a tiny minority of them to enter the upper reaches. Others have to make way for them by falling back down.

For someone to make any significant progress from the lower reaches to the higher requires super-human efforts and a lot of good fortune, quite different from what is required for someone born in the upper echelons to stay there. Those lower down the social order are generally required to conform to the expectations, demands and values of those above them if they want to climb higher. Insofar as this typically involves putting profits before people and human rights, before social solidarity and well-being, and before the environment, many people find the whole process ethically unacceptable and repugnant.

Positive incentives

Unrestricted inheritance makes a mockery of any defence of either just deserts or reward for effort in terms of necessary productive incentives because the really big rewards are handed out for no contribution whatsoever. Having been given away to the undeserving through inheritance, vast resources are therefore no longer available to provide 'incentives' for the deserving.

If the idea is the need to motivate necessary but unpleasant and unrewarding (perhaps dirty or dangerous) activities, then the reward should be in direct proportion to the unpleasantness in question. It is true that some dirty or dangerous jobs are quite well paid. Generally speaking the opposite is true: more rewarding and pleasant jobs are much better remunerated than unrewarding and unpleasant ones.

There is now a mass of evidence indicating that stress and anxiety contribute to increased mortality and morbidity, and that such stress and anxiety decrease higher up the job hierarchy (see Marmot, 2004). Those at the top not only have more intrinsically rewarding work, and much more pay, they also have less stress and longer, healthier lives.

Where training is available but expensive, the only way in which poorer people can pay for it is by getting loans, and remuneration sufficient to comfortably repay such loans could indeed be a significant motivation. But where the high cost of such acquisition means that only wealthy and privileged people can access them, then, clearly, final remuneration has no bearing upon such acquisition for most people. And poorer people, who could acquire and exercise the skills in question as effectively as more privileged people, never get a chance to do so.

The current hierarchical pattern of paid employment has little to do with individual free or rational choice to exert appropriate efforts, in light of proportional reward for effort and sacrifice, in a context of equal opportunity, and much to do with special privilege, and artificially restricted entry requirements. As the rich are assured of rewarding work, they never have to worry about doing the necessary but unpleasant and unrewarding jobs. They have little interest in providing positive incentives of monetary rewards proportional to such unpleasantness. They are quite happy to allow the threat of poverty and privation to drive poor people to do the jobs in question.

DISCUSSION TOPICS

1. Can the orthodox economic program be defended from the ethical and practical criticisms developed in this chapter?
2. If the criticisms are justified, what needs to be done to address the problems of the current system?
3. What are the implications of the analysis developed in this chapter for bioethics?
4. Competition drives productivity increases; more output for a given input. This means reduced demand for labour unless the economy grows fast enough. Unemployment means reduced consumption and a spiral of recession. Tax revenues decline as government welfare increases. National debt increases, with debt-servicing further undermining the economy. What are the implications of this need for growth in terms of resource depletion, pollution and climate change? Could the system exist without growth?

Health inequalities

This chapter looks at inequalities in health within and between nations, as well as changing patterns of health inequality through time. It begins to consider possible explanations for such inequalities in terms of social and economic forces and developments.

Inequalities in health between countries

As Hilary Graham explains in her incisive study of health and socio-economic inequalities, *Unequal Lives* (2007), most of the evidence showing how individual health is shaped by social circumstances comes from richer, developed countries. This is because of the lack of effective national data collection systems in poorer countries and the uneven distribution of funding for research into health outcomes, with less than 10% of all such funding by the public and private sectors devoted to investigation of the health of the poorer 90% of world population.

There is now considerably more, and more reliable, data on child health in low and middle income countries, than on adult health, with 'death rates among children under the age of 5' generally recognised as 'more reliable than other measures of population health' for such countries (Graham, 2007, p. 65).

Infant mortality is the chance of a child born in a specific year dying before the age of one year, and under five mortality – the chance of a child born in a specific year dying before the age of five. Both capture the scale of current global health inequalities when considered in relation to the World Bank's classification of national economies in terms of gross national income per capita, that is, the average value of goods and services produced by each of the country's citizens. (It is the gross domestic product (GDP) plus net income from other countries.)

Of the 12 million children under the age of five – 33 000 per day – dying in 2001, and more than 10 million in 2007, the great majority were in poorer countries (D. Roberts, 2008, p. 39). In countries with a gross national income (GNI) per capita greater than US$10 725 a year '7 in every 1000 children died before they reached the age of 5' in 2006. With a GNI of less than $875 a year in 2005 'over 120 children in every 1000 [did] not reach their fifth birthday' (Roberts, 2008, p. 24).

As David Roberts says, because of their undeveloped immune systems, children below the age of eight are particularly susceptible to water-borne diseases, 'that can kill when untreated'.

> Child survival is for the UN Human Development Index [therefore] one of the most sensitive pointers of human welfare, the comparative health of nations and the effectiveness of public policy. (2008, p. 38)

The UN has described the current situation as 'an international emergency' with 'roughly one child dying every 3 seconds in 2002' (D. Roberts, 2008, p. 38). An 'estimated 4 million die in the first month of life' even though most of the illnesses involved, including pneumonia, septicaemia, diarrhoea and tetanus (between them causing two in every three deaths), are easily 'preventable, treatable or both' through provision of 'the most basic health services' including cheap vaccinations and medicines (D. Roberts, 2008, p. 39). Lack of safe water plays a major role in the deaths of 1.5 million children from diarrhoea every year (Usdin, 2007, p. 10). Malaria kills one child every thirty seconds – 3000 per day under five years old, even though they could be protected by cheap netting and insecticide (Usdin, 2007, p. 40).

Many of these children – possibly five million per year – succumb to such diseases not just because of exposure to the relevant pathogens (in air, water and food) but also because inadequate nutrition leaves them especially vulnerable. They have inadequate food to 'fuel their immune systems to ward off diseases' (Usdin, 2007, p. 42).

There are some questions relating to interpretation of infant mortality figures. Some countries (including developing world countries and some European countries) apparently record the death of severely premature and low weight infants within minutes or hours of birth as foetal deaths rather than live births followed by infant deaths. Its been argued that major efforts in some countries, including particularly the US, to keep premature infants alive therefore increase infant mortality figures for such countries, creating a misleading impression of population health. Those arguing this case still acknowledge that it is relevant to small differences in overall rates only.

A more significant 'complication', both in terms of calculating the numbers involved and the explanation for these statistics, is that of infanticide, preferentially involving female children mainly in India and China, but also in other countries, including Libya and Turkey. As Roberts points out, some have argued that census data suggest as many as 60 or 100 million 'women and girls missing from normal population counts' around the world, including 'between 22 and 37 million Indian girls' mainly younger than four years (D. Roberts, 2008, p. 39).

These figures are contested and a significant percentage of 'missing' girls and women reflect medical abortions including dangerous backstreet operations, preferentially directed towards female foetuses. In other cases sex imbalances are the result of infanticide and selective neglect. Most important for present purposes, the evidence clearly indicates that such killings are concentrated in areas of greatest poverty and insecurity, with such poverty and insecurity functioning to maintain significant gender inequality and power imbalance. Women themselves are drawn into the process on the grounds of sparing female children from suffering in poverty-stricken patriarchal conditions (Roberts, 2008, pp. 34–5).

Female abortion and infanticide can be directly related to specific elements of corporate globalisation, including the Green Revolution (GR) in India (see Chapter 5), where 'introduction of industrial, chemical agriculture... displaced women from rural livelihoods' and thereby undermined their social power and value (Shiva, 2005, p. 136). Vandana Shiva (2005, p. 134, p. 135) observes that before the Green Revolution in India, child sex ratios were nearly in balance. After this, the ratio of boys to girls steadily increased. 'The GR region of Punjab' was the place where preferential abortion of female foetuses started. Increased demand for dowry from a girl's family to buy luxury consumer goods added to the pressures for killing female foetuses and female infants.

History

Data concerning the link between poor conditions and poor health have only been systematically collected since the mid nineteenth century and predominantly in the early industrialising nations of Northern Europe and North America. Records are particularly good from the UK, providing clear evidence of how health inequalities persist despite health improvements (Graham, 2007, p. 66).

In the early decades of the twentieth century there was a rapid decline in infectious diseases in high income countries. This spread to less developed regions, including parts of Asia, Africa and Latin America mid century, leading to a 'narrowing of global health inequalities' from the 1950s to the 1980s (Graham, 2007, pp. 66-7).

Life expectancy in Asia, the Middle East and North Africa continued to rise from the 1980s, moving closer to those in the high income countries of Europe and North America where the rate of increase remained constant or reduced. In sub-Saharan Africa, however, the increase stalled at a point where life expectancy was still under 50 years. This is a region where the HIV/AIDS epidemic is dragging down life expectancy. In South Africa, for example, life expectancy stood at 47 years in 2001. As Graham points out, it is a region where governments spend two to four times as much each year servicing debts as on providing education and health care (Graham, 2007, p. 67).

Global life expectancy increased by eight years in the two decades prior to 2001, reaching 66.7 years. On average there still remains a 20 year difference in life expectancy between low income countries – 59 years – and high income countries – 78 years. Death rates among children in the world's poorest countries are far higher than in the middle and high income countries, with one child in eight not living to their fifth birthday. Poorer, shorter-lived populations also suffer longer than average periods of ill health; for example, 18% of the average lifespan in sub-Saharan Africa as against 8% in Japan (Graham, 2007, p. 68).

In 1975, Samuel Preston demonstrated a strong relationship between a nation's wealth, as measured by GDP (the sum of all final expenditures on goods and services in an economy, including gross investment in fixed capital and net exports) per capita, and life expectancy. Recent data continue to support this finding: at low levels of GDP, below around US$5000 (adjusted for purchasing power parity), slight increases in GDP are significantly correlated with increased average life expectancy. For richer countries, such increases in GDP bring 'diminishing health benefits'.

> The marginal health gains of extra income are far greater at the lower end of the income range, meaning that redistribution of income from rich to poor would improve the average health outcomes of the latter far more than they would diminish the health outcomes of the former. (Graham, 2007, p. 68)

While national wealth, measured by GDP, and population health are strongly linked, there remain numerous outlier countries with lower or higher life expectances than their wealth would predict. The USA, in particular, had lower life expectancy (77 years) and much higher infant mortality in the early 2000's than countries with significantly lower GDP, including Sweden (with life expectancy of 79.9 years) and Japan (80.7 years) (Graham, 2007, p. 69).

Under-investment in employment protection and welfare provision in the US, compared to other advanced economies, keeps poverty rates (as measured by 50% of median income) and income inequality appreciably higher than in other rich countries. High rates of poverty and marked income inequalities exert a downward drag on life expectancy.

At the other end of the scale Sri Lanka has a GDP half of that of Brazil, but a higher life expectancy – 69 years for men and 76 years for women compared with 68 years for men and 75 years for women (in 2006, according to the WHO). Sri Lanka also has a lower infant mortality rate – 11 compared with 19 per 1000 (in 2006) – and a lower rate of under five mortality – 13 compared with 20 per 1000. Graham highlights Sri Lanka's 'equity-oriented approach to economic and social development' – with significant redistribution of income in the service of reduction of poverty and inequality – as the key to its success. This included substantial investment in public health and primary education from the early 1900s and food subsidies from the 1940s.

> Literacy rates are high – 90% – and over 75% [have] access to safe drinking water. Nearly 100% of pregnant women receive [quality] antenatal care . . . and over 80% of children receive [comprehensive] immunizations by their first birthday. (Graham, 2007, p. 61)

When Fidel Castro came to power in 1961, Cuba's mortality rates matched many other places in the developing world at the time, with a life expectancy of 48 for men and 54 for women. Infant mortality was 37.3 per 1000. Now Cuban statistics are comparable with those of Europe or the US. Life expectancy for both men and women is well up into the 70s (76.9 years in 2000). Infant mortality was 5.3 per 1000 in 2006 and 6.0 per thousand in 2008 compared with 7.2 per 1000 in the US. This was achieved at an annual cost of $186 per person (in 2000), around one twenty-fifth of the health spending in the US (Cooper, Kennelly and Ordunez-Garcia, 2006).

This indicates that 'the relationship between national wealth and population health is not a direct and simple one', but is 'mediated by government policies' with better health outcomes produced by income redistribution in the service of greater equality and welfare provision (Graham, 2007, p. 70). Such policies have been increasingly hard for middle and low income countries to sustain, in particular as a consequence of the structural adjustment program imposed by the World Bank and International Monetary Fund (IMF) as a condition of financial aid and debt rescheduling. As Graham says, these programs, by requiring the opening up of domestic markets to foreign investment, the cutting of public expenditure and privatisation of publicly provided services, 'have squeezed the "policy space" for national and sub national governments to invest in redistributive policies' (Graham, 2007, p. 71).

Inequalities in health within countries

Socio-economic inequalities within countries are measured by a range of indicators including occupational status, education and income. 'And using all such indicators, inequalities in health as a function of socioeconomic inequality are evident in both poorer and richer societies' (Graham, 2007, p. 72). For some health outcomes and for some age and gender groups, health inequalities take the form of a steady gradient with each step down the ladder of social inequality bringing a steady increase in the prevalence of poor health. The rates of ischemic heart disease – the leading cause of death amongst men and women in the UK – for men aged 35–64 between 1997 and 1999 show a steady increase from professional and managerial groups, with 90 deaths per 100 000 person years, to skilled non-manual, with around 120 deaths, skilled manual groups, with around 140 deaths, and unskilled and semi-skilled manual workers with 167 deaths.

In some poorer countries, like India (with a per capita GDP in 2003 of US$550), Nicaragua (US$750) and South Africa (US$3500), under five mortality rates show a sharp decline with rising household income. In all three countries, the rate is more than twice as great in the poorest quintile compared with the richest quintile (Graham, 2007, p. 73).

In other cases, the gradient flattens out at the higher and lower ends of the socio-economic hierarchy, demonstrating a 'threshold effect'. For example, during the 1990s in the UK the threshold effect was illustrated by families with children displaying specific mental disorders. For families earning above a certain level of gross weekly household income, increases in income were not associated with further declines in rates of mental disorders. However, for families earning below a certain level, there was a sharp increase in the rate of mental disorder with no further increases for the lowest income groups below this (Graham, 2007, p. 32).

Relative differences present health data in a way which shows how much more likely poorer groups are than richer groups to experience particular health problems. The data above suggest that the risk of ischemic heart disease for men in the poorest circumstances is nearly twice that of men in the most advantaged positions. Work by Kunst et al. compared socio-economic positions measured by education and health by self-assessed health in different European countries. The study compared the most advantaged groups with post-secondary education with the least advantaged with up to lower secondary education. In the UK the rates of poor health were 3.08 times higher among men and 2.66 times greater in women in poorer circumstances compared to those in better circumstances. The differences were less in the Netherlands with 2.81 and 2.12 and West Germany with 1.76 and 1.91, but greater in Austria with 3.22 and 2.67 for men and women respectively. Women in poorer circumstances were worse off in Finland

– 3.29, Sweden – 3.06, Denmark – 3.00 and Spain – 3.10 (Graham, 2007, p. 75).

As Graham observes 'wealth is unequally distributed in higher income societies', as in lower income societies but

> poorer people fare better in some countries than others... The USA has much higher GDP per person than the UK, but socio-economic disadvantage is associated with much higher rates of ill health in the richer USA than the poorer UK. (2007, p. 75)

She cites diabetes rates twice as high amongst the most disadvantaged 55 to 64 year olds in the US at more than 14% compared with 7% in the UK. She also highlights the higher rate of 'premature death for a male manual worker between 45 and 65' in Ireland (29%) as compared to Sweden or Norway (20%) in 1998 (Graham, 2007, p. 75).

More unequal societies are more unhealthy

As Wilkinson points out, 'the quality of social relations – as indicated by levels of trust, violence and involvement in community life', is enhanced 'in societies with smaller income differences between rich and poor'. Due to the importance of such considerations in determining health outcomes, such outcomes are 'likely to be worse in more unequal societies'. In such societies, health

> is compromised not only by the bigger burden of low social status and relative deprivation that goes with greater inequality, but also by the poorer quality of social relations. (Wilkinson, 2005, p. 101)

There are a number of common ways of expressing differences in the extent of income inequality. Researchers sometimes use the proportion of the society's total income received by the poorest half of the population, which is in many cases around 20%, compared to 80% received by the richest half. Sometimes they consider what proportion of total income would need to be redistributed from rich to poor to equalise all incomes (called the Robin Hood Index).

> The most common measure is... the Gini Coefficient... [which] runs between 0 (everyone gets the same amount) and 1 (all income goes to one person.) Values for many societies vary around 0.3 or 0.4. [Most] measures... tend to be closely correlated..., so... choice of measure usually makes [little] difference to the results of analysis of the relationships between health and inequality across... societies. (Wilkinson, 2005, p. 102)

Wilkinson's own recent research has used a scale based on how much more (after tax and benefit) household income is received by the richest 20%, than by the poorest 20%, in each country. According to this scale, the richest 20% of the US receive more than eight times more income than the poorest 20%, the richest 20% in the UK and in Australia around seven times more; in Sweden and Japan less than four times more. (Wilkinson and Pickett, 2009, p. 17)

We have already noted that infant mortality rates in developing countries decrease with increasing gross national income (GNI) per capita. The figures show infant mortality coming down rapidly with increasing GNI per capita amongst the poorer countries before progressively levelling out in richer countries. Among the richest countries further economic growth makes little or no difference to mortality. Robert Waldman, using World Bank data on income inequality for 70 different countries around 1970, found that after controlling for gross national product (GNP) (very close to GNI) per capita, infant death rates were higher for more unequal countries. This conclusion has been confirmed by Hales and colleagues using more recent data from the 1990s. Their work shows that at all levels of economic development infant mortality rates tend to be lower in more egalitarian societies.

Amongst richer countries there is no longer any strong relation between higher GNP and better health outcomes. The extent of inequality becomes the most significant determinant of health outcomes. As Wilkinson points out, there is little relation between even twofold differences in average income and life expectancy in the developed world, including the 50 US states. However, within countries and states

> there are large health inequalities related to relative income and social status. As the effects of absolute poverty have weakened, the social effects of relative deprivation have been unmasked... (Wilkinson, 2005, p. 105)

Kennedy et al. looked at the results of numerous studies of the relations between income distribution and age adjusted death rates in the 50 states of the US. They found clear evidence that it is the most egalitarian states (measured in terms of total income received by the poorest 50%) that are the healthiest. Death rates (among men 25 to 64 years old) are nearly twice as high (675 plus per 100 000 people) in the most unequal states (with less than 18% of income going to the lowest half of the population) compared to the least unequal (with more than 23% of income going to the lowest half of the population). 'Although the relationships between inequality and health have been shown at all ages, they seem to be strongest among men of working age' (Wilkinson, 2005, p. 105).

Going beyond North America, Nancy Ross and Jim Dunn have put together data from 528 cities in five different countries for which data were available on a comparable basis: the US, the UK, Sweden, Canada and

Australia. They found a striking tendency for death rates to be higher in cities where there is more inequality. 'The relationship appears consistent across all the cities from the most unhealthy and unequal American ones to the most egalitarian Swedish...cities' (Wilkinson, 2005, p. 109).

A recent analysis of data from the 20 regions of Italy has found 'a close relationship between the extent of income inequality and average life expectancy in each region' (Wilkinson, 2005, p. 109).

Using the richest/poorest 20% income scale noted earlier, Wilkinson and Pickett find a strong correlation between income inequality in developed world countries and life expectancy. In Japan and Sweden average life expectancy is over 80 years. In the UK it is around 78 years and in the US around 77 (Wilkinson and Pickett, 2009, p. 82).

In Sweden and Japan, infant mortality is around 3 deaths per 1000 live births, while the figure is around 5.5 in the UK and close to 7 in the US. Similarly, around 10% of adults and children are obese in Sweden, but more than 20% of adults and 15% of children in the UK and more than 30% of adults and 25% of children in the US (Wilkinson and Pickett, 2009, pp. 92–3).

Women

We have already touched upon the issue of female infants being principal victims of infanticide in some poorer countries. There is substantial evidence of older girls and adult women also suffering significantly worse health outcomes (than boys and men), particularly amongst the poorer people of some developing world countries.

Usdin estimates that 70% of the '1.2 billion people living in extreme poverty are women' (Usdin, 2007, p. 78). This has profound consequences for health outcomes, including denial of basic education in health matters. The United Nations Development Programme (UNDP) finds that 'the maternal mortality ratio is 540 deaths per 100 000 live births' for much of Africa and Asia, compared with a much lower ratio in the developed world, while also noting that complete absence of data from 62 countries means that this is an underestimate (D. Roberts, 2008, p. 44). It notes that three quarters of deaths of pregnant women could be avoided with low cost interventions – including better education and access to trained medical staff.

Around the world, 'over 120 million women and girls are estimated to have undergone genital mutilation or cutting' with various degrees of damage to their sexual organs 'and 2 million more undergo the procedure every year' (Usdin, 2007, p. 76). As Usdin says, this has a range of serious health consequences in addition to pain, shock and blood loss, which include

infections [of] the urinary tract and pelvic inflammatory disease, damage to
the urethra leading to incontinence, HIV transmission from shared
instruments, menstrual problems, infertility, painful... intercourse... or lack
of sexual fulfilment with subsequent psychological [consequences]. Many
women report that husbands have other sexual partners as a result, placing
them at further risk of HIV and other STDs. (Usdin, 2007, p. 77)

Women are victims of domestic violence, sexual violence, dowry murders and honour killings. In some parts of the world the murder of women by their male partners finds wide social acceptance (70% of all women murdered globally according to Amnesty International) (D. Roberts, 2008, p. 46). David Roberts (2008, p. 59) cites Indian National Crime Records of one woman killed every day because of 'insufficient dowry', with 'other marriage killings' happening 'approximately once per hour'.

Women are killed by male relatives in Pakistan, Egypt, Yemen and many other countries, because of 'immoral behaviour' including refusing to submit to arranged marriages, marital infidelity, failing to serve meals on time or 'allowing themselves' to be raped (D. Roberts, 2008, p. 55). In Europe

> domestic violence against women is the major cause of death and disability
> for women aged 16–44 and accounts for more death and disability than cancer
> and traffic accidents... Amnesty... declares that in the USA, 4 women die
> each day as a result of violence in the family [and] in Russia, 14 000 women
> were killed by an intimate partner or ex-partner in 1998. (D. Roberts, 2008,
> p. 47)

As Usdin (2007, pp. 74–5) points out, women are particularly vulnerable to HIV 'because the cervix is receptive' and sperm sits for long periods in the vagina, and 'violence against women places them at greater risk of traumatic sex, with bleeding increasing the likelihood of HIV transmission'.

Usdin also notes that

> 2 million girls between the ages of 5 and 15 are... forced... through
> poverty... into the sex market each year... Many are... kidnapped while
> others are sold by desperately poor families or families that undervalue the girl
> child. In this modern form of slavery, they face violence and rape and are
> vulnerable to STDs including HIV. Most are denied even basic health care.
> (2007, p. 73)

With tobacco companies specifically targeting women and girls, the death rates from smoking related diseases amongst women have increased in recent years, to the point where cigarettes are now

> responsible for 30% of all middle-aged female deaths... Additional
> risks... include infertility, premature labour, cancer of the cervix and
> osteoporosis. Smoking while on oral contraceptives is associated with a
> tenfold increase in heart disease. (Usdin, 2007, p. 113)

Smoking mothers can expose their infants and children to side-stream tobacco smoke. The World Health Organization (WHO) estimates that 250 million of the 700 million children passively smoking today 'will die from tobacco related diseases' (Usdin, 2007, pp. 111–12).

Health inequalities over time

As Graham explains, the industrial revolution in Britain saw a rapid expansion of urban populations, particularly in the north of the country after 1800. Here, poor sanitation and overcrowding contributed to high levels of 'tuberculosis, cholera and diphtheria, along with infant diarrhoea', bringing average national life expectancy down to around 40 years and pushing infant mortality up to 460 per 1000 in under fives in industrial Liverpool in the 1860s compared with 175 per 1000 in prosperous Hampstead (Graham, 2007, p. 77). In 1838–41, the average age of death of the gentry and professional groups in English cities was over 40 years while that of labourers and artisans was around 20 years.

> From the 1870s survival rates improved and life expectancy started to rise [with a] decline in mortality from communicable diseases. At first the rise... was modest, climbing from 41 years in 1870 to 47 years in 1900. Across the early decades of the twentieth century the upward trend was steeper [reaching 58 years by 1921 and 65 years by 1951]. The improvement in life expectancy among the non-combatant population was particularly marked during the war decades of 1910–20 [+6years] and 1940–50 [+6 years]. (Graham, 2007, p. 78)

Improving living conditions for the urban poor, including rising real wages and environmental improvements in public health, housing, factory safety and welfare services, seem to have played a major role in initial improvements.

Graham highlights the provision of 'integrated sewerage systems and piped clean water, improved housing conditions and improvements in the quality and regulation of the food supply' as playing a crucial role in driving significant health gains in the period from 1870–1910. She implicates such 'wartime welfare measures' as price controls and equal access to basic necessities, including food subsidies and rationing, as improving conditions for working class families in the 1940s.

> At the same time [government] extended welfare benefits and strengthened people's legal entitlement to health care... Tax changes further contributed to a levelling up of living standards of poorer households while holding down incomes of richer households. (2007, p. 78)

Richard Wilkinson (2005, p. 113) highlights the significance of the disappearance of unemployment during the Second World War, along with the 'dramatic reduction in income differentials' as a result of effective government takeover of the economy. He sees these as central to the causation of psychosocial forces which played a key role in contributing to wartime improvements in life expectancy 'well over twice as fast as the average rate of improvement during the rest of the century'.

Reduced unemployment and reduced income inequality combined with a 'psychological sense of unity in the face of a common enemy' and with deliberate government policy to foster feelings of 'social unity and cooperation' to produce a widespread 'sense of camaraderie, of people pulling together and a sense of social cohesion' (Wilkinson, 2005, p. 115). Such psychosocial developments boosted immunity and improved health outcomes.

As Graham emphasises, historical analysis of this century of health improvements in the UK 'suggests that advances in medical care made a relatively small contribution to the rapid decline in mortality in the late nineteenth and early twentieth centuries'. Though 'from the 1950s medical care played a larger role in improving health' with 'the middle decades of the twentieth century' seeing 'major advances in the treatment of infectious and chronic diseases'.

> They saw, too, the expansion of health care systems in many high income societies, funded through taxation – as in the UK – or by compulsory payroll deductions [which] widened access to quality care. (Graham, 2007, pp. 78–9)

The UK's National Health Service (NHS), launched in 1948, aimed to provide comprehensive medical, dental and nursing care for all 'rich or poor, man, woman or child'.

From a psychosocial perspective, the development of such comprehensive public health systems can be seen to have reduced stresses and fears of illness, a lack of access to care and accumulation of unpayable debt to pay for private health care on the part of poorer people. Such stresses and fears themselves contributed to ill health so that public health provision probably contributed to improved public health even for those not accessing such services.

The epidemiological transition

The shift from an agrarian economy to one based on manufacturing and service industries has been associated with changes in the major causes of death from communicable diseases like TB, malaria, cholera, yellow fever (along with accidents and malnutrition) to non-communicable diseases. As Graham points out, these non-communicable diseases, including ischemic heart disease, cerebrovascular disease, malignant neoplasms of the trachea, bronchus and lung, chronic lower respiratory disease, dementia and Alzheimer's disease, typically develop through long-term exposure to risk factors. In part, their increasing importance is a consequence of the decline in child mortality and the increasing proportion of the population living long enough to die from such slowly developing conditions.

Most importantly, it reflects a rapid increase in deaths from circulatory disease and cancer across a period in which the death toll from infectious diseases was falling rapidly. Figures from the UK, for example, show a steady decline in deaths from infectious diseases from 1911-71 (after which they have remained at a stable very low level), while cancer deaths steadily increased from 1911-91, overtaking infectious diseases around 1931. Ischemic heart disease deaths overtook infections around 1916 and rose steadily till around 1971, falling off after that, but still remaining at a high level.

The concept of an 'epidemiological transition' was coined by Abdel Omran in the 1970s to describe the change in major causes of mortality in early industrialising countries of Northern Europe. He called it

> a long term shift in mortality and disease patterns whereby the pandemics of infection are gradually displaced by degenerative and man-made diseases as the chief form of morbidity and primary cause of death. (Graham, 2007, p. 84)

Evidence shows that while infectious diseases remain the major cause of child mortality in low to middle income countries which have undergone recent industrialisation

> non-communicable diseases now account for more than 50% of adult deaths in all regions except south Asia and sub-Saharan Africa. In low, middle and high income countries, ischemic heart disease and cerebro-vascular disease are the leading causes of adult deaths. (Graham, 2007, p. 84)

Usdin (2007, p. 110) notes that 'the WHO estimates that 80% of the 35 million annual deaths due to non-communicable diseases in 2005 occurred in low and middle income countries'.

Particularly in countries where HIV/AIDs and malaria kill large numbers, the transition is leading to a 'triple burden' of mortality, with high rates of acute infectious diseases persisting alongside increasing chronic, non-infectious disease and high levels of mortality from road traffic accidents and violence (Graham, 2007, p. 85).

As Usdin says, low birthweight due to poverty can contribute to chronic disease, but Western 'food, alcohol and tobacco' industries bear a major responsibility for the active promotion of dangerous products and lifestyles in the developing world, as do manufacturers, exporting high pollution industries, that 'cause asthma and lung disease' (Usdin, 2007, p. 110).

In the developed world, by contrast, death rates from chronic diseases have been dropping, but the rate of decline has been faster among more advantaged groups, leading to widening health inequalities. A key factor here has been a more rapid reduction in risk factors for ischemic heart disease and lung cancer, including cigarette consumption, high fat diets, physical inactivity, overweight and obesity amongst more privileged groups. National rates of overweight and obesity have continued to increase in both developed and developing countries, including the UK (with obesity doubling between 1980 and 2000), Australia, the US, parts of the Middle East, North Africa and Latin America. Childhood obesity has increased in the US, China and South Africa.

> WHO estimates that in 2005, 300 million people were obese and over 1 billion were overweight... at least 185 million school age children classified as overweight... 30–45 million as obese. (Usdin, 2007, p. 118)

As Usdin maintains, processed food corporations around the world have played a major role in promoting foods high in 'sugar, calories, saturated fat, trans fatty acids and sodium' (2007, p. 118) spending 'US$30 billion on direct advertising and promotion in the US alone' (2007, p. 119). This issue is discussed in Chapter 5.

The role of class structure

Dahlgren and Whitehead talk of the main determinants of health inequalities as 'a series of layers, one on top of the other' (1991, p. 11; Graham, 2007, p. 103). Globalised economic and cultural forces and processes provide the broad base, shaping the socio-economic conditions of particular nations and regions. Such conditions, in turn, shape individual living and working arrangements, including wage rates, 'housing, education and healthcare'.

> Mutual support from family, friends, neighbours and the local community comes next. Finally there are actions taken by individuals, such as the food they choose to eat, their smoking and drinking habits. The age, sex and genetic make up of each individual also play a part... (Graham, 2007, p. 103)

Causation is seen to operate from the 'outer', environmental layer, to the inner, individual layer. Societal level factors of general socio-economic, cultural and environmental conditions 'exert their influence on a set of middle range determinants, labelled 'living and working conditions' and 'social position'.

> Health behaviours – lifestyle factors and specific exposure – are placed further along the causal pathway... as the last of the social influences on health. From this point, genetic and biological factors take over. (Graham, 2007, p. 106)

Socio-economic structures and processes at the national and international level assign individuals and families – and regions – to particular social class positions. At the core of such positions are radical disparities of power over the production and distribution of material goods, services and ideas, and over the political process. Such significant disparities of social power are closely correlated with disparities in wealth, income, education and living conditions, and health outcomes.

There will always be class division and class inequality in a capitalist society that deprives the majority of effective ownership and control of major productive resources and forces them to work for those who exercise such control. But the nature and consequences of such division and inequality are significantly influenced by dominant political structures and policies. There can be significant differences in the life circumstances – including the health – of working class people. This is dependent upon whether and to what extent capitalist class inequality is moderated by social-liberal policies of effective redistribution and welfare provision, through taxation and through direction of the profits of state controlled industries, along with active democratic engagement of the mass of the population. In particular, such life circumstances and health outcomes can be significantly improved by policies of guaranteed full employment, workers' representation and living wages, by public provision of comprehensive health care, education and housing, through support for trade union organisations and collective bargaining, and for increasing collective workers' control of working conditions.

The developing world

The role of class structure and class inequality in shaping health outcomes is reasonably clear in the developing world. While the rich and the powerful can be as rich and powerful as rich and powerful people anywhere, the poor are generally desperately poor and powerless compared to the poor of the more developed world. There is little in the way of moderation of capitalist class inequality by any kind of redistribution and welfare provision. As will be explored in detail in later chapters, in the less developed world, poverty and powerlessness increasingly mean limited or no access to clean water and healthy food, housing or sanitation.

There is a close relationship between good nutrition and good health. If children do not absorb and retain sufficient calories, vitamins and protein due to diarrhoea and/or measles early in life, their immune systems are weakened, leaving them vulnerable to other diseases (Mehrotra and Delamonica, 2007, p. 104). Furthermore, micronutrient shortages and illness can radically impair development, with iron deficiency reducing cognitive functions, iodine deficiency causing 'irreversible mental retardation' and vitamin A deficiency 'the primary cause of blindness among children'.

Safe water and adequate sanitation are also crucial in maintaining good health outcomes. 'Access to safe water and sanitation dramatically reduces the incidence of diarrhoea and many other diseases that kill millions of children and adults each year' (Mehrotra and Delamonica, 2007, p. 105). With water on tap, women and girls have time free from long distance water transport to devote to better infant and child care. Also 'lower infant and child mortality plays a major role in reducing fertility rates' (Mehrotra and Delamonica, 2007, p. 104) and thereby reducing pressures of population and poverty.

In the developing world poverty and powerlessness mean often appalling working conditions, involving long hours of backbreaking toil, often from early childhood, for below subsistence wages, often exposure to high levels of toxic fumes, vapours or dusts, without any proper health or safety protection.

On big plantations, workers frequently toil from dawn till dusk, for US$2 a day or less, bullied and intimidated by armed overseers, and bathed in the toxic chemicals of industrial agriculture – insecticides, fungicides, herbicides and fertilisers. As Weis (2007, p. 112) says, 'the everyday toxicity facing estate workers of high-input modern farms poses risks of short-term poisonings as well as long-term health implications'. He refers to WHO figures of 250 000 deaths each year from acute pesticide poisoning (2007, p. 31).

Developing world miners are at particularly high risk with '300 large scale accidents in illegal mines along the West Bengal–Jharkhand coal belt' alone 'claiming the lives of at least 2000 miners' every year (Moody, 2007, p. 73). While increasing numbers of workers in the industrialised world are 'now dying from past exposure to asbestos dust... millions in the global south, especially in China, India, Thailand and Indonesia face daily contact with these carcinogenic fibres in their homes, factories, schools and public buildings', pushing asbestos fatalities above 100 000 per year (Moody, 2007, p. 76). India, in particular, is increasing its use of asbestos in cement, as other countries have banned it (Moody, 2007, p. 114). Thirty seven per cent of miners in Latin America, possibly half a million in South Africa and 10 000 more people each year in China have 'some degree' of silicosis, which could now have overtaken asbestos as the major 'industrial killer' (Moody, 2007, p. 78).

Mining and mineral extraction operations around the world, particularly metal refining and smelting, including wastes from past operations which have not been cleaned up, continue to affect not only miners and mineral workers but also surrounding populations. These people cannot afford to move to safer areas, so are exposed to toxic gases and dusts, volatile organic compounds, radioactive isotopes, hexavalent chromium and heavy metals, including lead, cadmium, copper and cobalt, in the air, water and local foodstuffs.

Roger Moody identifies a polymetallic smelter in La Oroya, Peru, a chromium factory in Ranipet, India, a closed down lead smelter in Haina,

Dominican Republic, past mining and smelting in Kabwe, Zambia, radioactive uranium mine tailings in Mailuu-Suu, Kyrgyzstan, and coal mining and steel smelting in Linfen, Shanxi Province of China, (amongst others), as major 'sacrifice areas' responsible for undermining the health of hundreds of thousands of people (Moody, 2007, pp. 135-43). He highlights the continued toxic pollution of entire river systems by the Grasberg mine in West Papua and the Ok Tedi mine in Papua New Guinea.

Millions in the developing world have no access to employment at all, or to any compensating welfare facilities, and are forced to live in conditions of extreme poverty and misery.

Poverty and powerlessness mean limited access to education, and therefore to the huge health care benefits of education. 'Parents . . . . make better use of information and reproductive health care facilities if they are more educated' (Mehrotra and Delamonica, 2007, p. 102). So less education is associated with higher fertility. Less educated parents provide worse nutritional and health care for themselves and their children.

> The general knowledge acquired at school increases understanding of modern health practices . . . educated mothers [spend] a higher proportion of their income on food and health services . . . the capacity to acquire new knowledge and change behaviour accordingly is higher amongst those who attended school, as evidenced by the differential diffusion of HIV/AIDS amongst educated and uneducated women . . . (Mehrotra and Delamonica, 2007, pp. 102–3)

Low levels of education coupled with high levels of stress – and of advertising – render poor people particularly vulnerable to addiction to cigarettes (and other drugs) and junk food, and use of worthless and dangerous pharmaceutical products and other 'treatments'. In the absence of public provision, poverty denies them access to private health care, and often such care as is available, is hopelessly inadequate.

Obviously governments in the developing world have limited capacities for addressing these issues, compared to those of the developed world. We need to understand why this is the case, why potentially wealthy countries find themselves in such dire straits. This involves looking at their relations with the countries of the developed world. See Chapter 3 for some relevant issues.

It is not simply a question of limited resources. We have already noted how some developing countries and provinces manage to achieve good health outcomes despite their limited resources. Despite the often very low levels of GDP of developing compared to developed countries, effective redistribution and sharing along with rational planning and direction of resources can still allow good population health outcomes. Health care costs can be kept down by effectively addressing the social causes of ill health.

An important example here is provided by redistribution and inequality reduction in the Indian state of Kerala. With only about a hundredth of the per capita income of richer countries and spending only US$28 per capita per year on health compared to nearly US$4000 in the US, it maintains life expectancies of 76 for women and 74 for men, roughly on a par with those of the US (Usdin, 2007, pp. 14–15). Kerala loses 14 infants per 1000 live births, which, while higher than the overall US rate, is a lower figure than for African Americans (Usdin, 2007, p. 15). As Usdin points out, this is achieved through combining 'basic curative and preventative health services with strategies to ensure land reform, universal access to housing, education (emphasising gender equity), subsidised school transport and nutrition, water, sanitation and extensive social safety nets' to reduce inequality.

Unfortunately, as we have already seen in the last chapter, there remain strong internal and external pressures upon poorer countries working against any such redistribution and rational planning.

The developed world

Pockets of developing world-level poverty and deprivation persist in particular areas of the developed world, including substantial numbers of indigenous peoples and 'illegal migrants', deprived of basic necessities of clean water and sanitation, healthy food, housing, education and health care. Outside such groups comparatively few people are without access to clean water, a basic diet and a roof over their heads. Most have access to some sort of social security support in case of unemployment, including basic health care. While there are still plenty of serious hazards in the working environments, particularly of the less skilled and manual workers, including ongoing exposure to toxins and carcinogens, nonetheless, there are generally qualitatively greater legal protections than exist in the developing world.

Yet significant social class differences in health outcomes still persist, with five, 10 or sometimes 15 years' difference in life expectancy between rich and poor within the same society. Certainly, as Graham points out, 'lifestyle' factors, particularly cigarette consumption and eating habits, play a role here. By the 1950s, with smoking a widespread habit among men and women in the UK, its prevalence in higher socio-economic groups was already falling. With increasing dissemination of the health risks of smoking, the downward trend continued through the 1960s and early 1970s. However, reductions were much less significant among poorer people. While the reductions have continued since the 70s, they have been 'modest' amongst poorer people, even showing increases amongst working class women after 1998 (Graham, 2007, pp. 91–2).

As Usdin (2007, p. 111) notes, 'big tobacco spent over US$15.2 billion in 2003 on marketing and promotions in the US' despite increasing legal restrictions over recent decades. Tobacco related illnesses remain 'the leading cause' of preventable adult death in the northern hemisphere.

Similarly, while 'socio-economic inequalities in adult obesity were largely absent in the 1960s, socio-economic gradients were already beginning to emerge for girls' with 'the upward trend in obesity and being overweight more marked among children from poorer backgrounds'. Such gradients later emerged for boys and for adults while 'steepening' amongst girls (Usdin, 2007, p. 93). Graham (2007, p. 93) says, 'as rates of adult obesity and being overweight have risen, the association between socioeconomic disadvantage and excess body weight has strengthened'.

We still need to explain such class differences. So too do we need to explain the persistence of health differences across the social class spectrum even when controlling for such risk factors as smoking, diet, exercise and other relevant behaviours, as demonstrated by Marmot and others. Substantial amounts of research from around the world point to the importance of psychosocial welfare and the quality of the social environment in determining health outcomes in developed societies. In particular, they highlight the significance of increasing stress down the social power hierarchy as a key factor influencing health outcomes in these societies.

Marmot's classic studies of different employment grades in the occupational hierarchy of the British public service demonstrated a direct relation between ranking, mortality and morbidity. As he says, 'The men at the bottom of the hierarchy [had], at ages 40 to 64, four times the risk of death of the administrators at the top of the hierarchy' (Marmot, 2004, pp. 38–9). Each occupational level was found to have higher mortality than the one above it in the hierarchy. The gradient got shallower at older ages, 'but even at the oldest age, the bottom group [had] twice the mortality of the top group.' Those lower down the hierarchy had higher levels of lifestyle risk factors, in the areas of smoking, blood pressure, plasma cholesterol and blood sugar which did contribute to higher mortality and morbidity. But adjusting for such factors made only a slight difference to the results. In other words, smoking reduced life expectancy at the top and the bottom, but smokers at the top were still doing much better than smokers at the bottom. Something was protecting them (Marmot, 2004, pp. 44–5).

Since the gradient applies to incidence of disease and since relatively high quality public health services were, at the time, available to all British citizens, quality of health care does not explain the difference. Those lower in the hierarchy were found to have much more health care by virtue of higher levels of serious illness.

Evidence from many different directions, particularly experimental manipulation of status hierarchies in monkey populations, as well as large-scale epidemiological studies of human populations, point to work stress as the major cause of both the underlying health gradient and of the greater lifestyle risk factors down the hierarchy.

Primate experiments carried out at Bowman Gray University in North Carolina, with monkeys fed diets high in saturated fats and cholesterol, produced particularly suggestive results. With such monkeys 'naturally' forming themselves into hierarchies, it was found that the higher the rank of the individual, the less likely they were to develop atherosclerosis which is the underlying problem in coronary heart disease. In a 'changing ranks' experiment the top two monkeys from 'each of a number of groups' are put together in a new group. These animals are found to form themselves 'into a new hierarchy'. The same applies to previously low ranked individuals put together in new groups; they too form new hierarchies.

> In the experiment then, some low-rank animals have become high-rank and some high-rank animals are lower in the hierarchy. It is the new position, not the one they started with, that determines the degree of atherosclerosis that the animal develops. (Marmot, 2004, p. 119)

So it is not the case that some animals have a genetic predisposition to both dominance and good health, or subservience and bad. It is the social situation that determines health outcomes, with a high status protecting individuals from risk factors including poor diet. The same seems to apply in human society.

As Marmot says, 'the nature of the hierarchy is the less control' and less reward for effort 'the lower you go' (2004, pp. 129–30). A 'fight and flight' response, evolved to deal with brief emergencies, becomes increasingly chronically activated by increasing insecurity, powerlessness and fear as we move down the social hierarchy. Without the opportunity to relieve stress through fight or flight (because poorer people have to hold on to their jobs), the prolongation of the stress reaction of increased heart rate and blood pressure and rapid transmission of nutrients to muscles through the inhibition of insulin secretion undermines the bodily functions of tissue repair, growth, immune response and digestion, lead ultimately to illness and death.

Lower ranking public servants, at the time of research, had job security and living wages as well as access to comprehensive social welfare provision. Those in the lower reaches of corporate hierarchies in the contemporary world typically lack all of these things. They are in a situation where job loss will have a potentially disastrous impact on all aspects of their lives. This is in radical contrast to those at the top who have the financial and social resources to see them through job loss without any such consequences.

It's also important to consider the ways in which increasing inequality impacts upon the consumption patterns of those on the lower rungs of the social hierarchy through the mediation of powerlessness and chronic anxiety. Folk psychology tells us that anxiety, stress and powerlessness contribute to increased consumption – and reduced capacity to break free from – psychoactive drugs including alcohol, tobacco and caffeine, as well as the various legal and illegal stimulants and anti-anxiety and anti-depressant drugs. We know that stress also increases our consumption of sweet and fatty foods.

Massive corporate advertising, marketing and the intrinsically addictive characteristics of some key health damaging substances are important. Equally important is the susceptibility of consumers to this type of manipulation and addiction. As Wilkinson reminds us, we consume cigarettes and alcohol and cannabis to 'ease social tensions' and 'relax'. Coffee and amphetamines prime us to continue the struggle, Prozac lifts our depression and cheers us up, Valium and Zoloft calm us down and reduce our anxiety. Similarly, sweet and fatty foods seem to provide comfort in face of stress, unhappiness and unmet needs.

Wilkinson points out that these everyday intuitions and observations are supported by quantitative empirical research, which indicates, for example, increased cigarette consumption and obesity amongst the poorest quarter of the population of the UK at a time of increasing inequality and job loss in the later 1980s. While the poor have just as strong a desire to give up smoking as the rich, their social situation makes it much more difficult for them to do so.

DISCUSSION TOPICS

1 What general conclusions can be drawn from the evidence presented in this chapter about the nature and causation of individual ill health?
2 What can or should be done to address the issues of current health inequalities raised in this chapter?

Food and water

Corporate 'free' markets, with competition driving ongoing innovation, production where costs are lowest and transport to where monetarily effective demand is highest, have had particularly problematic consequences in the area of food production and distribution. Here we see a vicious contradiction between private property and need, with more than enough food for everyone today, but significant numbers of people going hungry because they have no money or power sufficient to access such resources, and the threat of many more going hungry in the future.

In his book, *The End of Food* (2008, p. xii), Paul Roberts describes a system where large scale livestock operations and chemically intensive grain production have 'so degraded the productive capacity of our natural systems' that it is far from clear 'how we'll feed the [9 to] ten billion people expected by mid century'. This is a system making a massive contribution to pollution and global warming while leaving nearly a billion people in the developing world 'food insecure', meaning seriously malnourished, a number increasing by 7.5 million per year. At the same time, increasing numbers of people in the developed world suffer diet related conditions of obesity, heart disease and diabetes and become increasingly obsessed with diets and food fads while food-borne infections including *E. coli* and salmonella become increasingly difficult to control.

This chapter draws upon the work of P. Roberts, Weis, Bello, Shiva and others to explore the current state of food and water provision around the world, and its consequences for public health.

Past developments

Roberts shows how the general problems of the market, outlined in Chapter 3, are expressed in particularly pure and insidious form in food production. Capitalist competition has contributed to increased productivity through labour-saving technologies in farming, as in other areas of production. 'Farms [have] come to be run like integrated factories turning 'inputs' of seed, feed and chemicals into standard outputs of grain and meat' (P. Roberts, 2008, p. xiii). At the same time, small-scale retailers have been increasingly replaced by chains of large-scale supermarkets (e.g. Wal-Mart, Tesco, Carrefour, Metro, Kroger), using their 'enormous volumes and market share to squeeze discounts from food companies'. This, in turn, has increased the pressures upon farmers to cut costs through increasing technology and scale economies. This has led to increased concentration and centralisation of farming activities with the disappearance of small family farms and increasing oversupply.

As noted in Chapter 3, without monopolisation, capitalist competition enforces reduced production costs through increased mechanisation. With widespread adoption of more productive, cost-cutting technology, the danger of overproduction increases. As P. Roberts explains, a big new combine harvester is a major expense at US$400 000

> but because it lets the farmer produce more wheat [they] can spread the cost over more bushels. Of course...all...farmers everywhere else in the industrialised world are also trying to spread their costs over more bushels. (2008, p. xv)

This led to half a century of supply rising faster than demand in the developed world particularly the US, leading to continuous price falls, which, in turn, 'forced farmers to invest in more technology to produce still more bushels and so on' (P. Roberts, 2008, p. xv). As P. Roberts notes, this cost-cutting treadmill now includes everyone in the supply chain from fertiliser producers to feed sellers, all striving to reduce their per unit costs. It goes a long way towards explaining both 'the aggressive way the United States and other surplus producers push their surpluses into foreign markets' through multilateral and bilateral free trade agreements, with disastrous consequences for producers in the developing world, and the epidemic of overconsumption in the developed world, with the proliferation of 'fast food' and 'value meals' (P. Roberts, 2008, p. xv).

Neoliberals and corporate capitalists in the West have pressed hard for corporate 'free' markets in foodstuffs on a world scale. The neoliberal argument in favour of such markets is that of comparative advantage, with universal benefit from increased productivity and lower cost over-riding any possible benefits of self-sufficiency. The increasing specialisation and division of world food producing labour in the name of 'comparative advantage' with some regions and nations specialising in single crops (of coffee, cocoa, sugar) or depending mainly on food imports has left huge areas massively vulnerable to inevitable future disruptions of production, sale and supply.

Poor people are increasingly reliant upon imports of basic foodstuffs. Those with money to spend are targeted with massive advertising of junk foods, as the US and European food processing industries intensify their marketing operations in the developing world. Meanwhile, the rich get foodstuffs from all around the world, with increasing access to cleaner and safer organic products.

Increasing scale and complexity of the supply chain, along with massive use of antibiotics in farm animals leading to antibiotic resistant strains of food borne pathogens, and continued pressure from business for deregulation and free trade, make it 'harder and harder' for governments to operate 'robust food safety programmes' (P. Roberts, 2008, p. xvi).

Roberts highlights the radical failure of the current system to eliminate hunger and malnutrition. UN figures indicate that while for some time there has been something like one and a half times as much food being produced as is required to provide a reasonable diet for the whole world population, nonetheless, as noted above, close to a billion people, mostly in the developing world, are undernourished

> Beyond the 36 million hunger related deaths each year, chronic malnutrition freezes entire populations in a [state] of constant exhaustion, stunted bodies and ravaged potential. By one estimate, the lack of a simple micronutrient, vitamin A, has left more than 3 million children under the age of 5 in sub-Saharan Africa permanently blind. (P. Roberts, 2008, pp. xvi–xvii)

While food costs fell for decades and supply increased (far beyond existing food needs), the number of people unable to buy such food continued to increase. The logic of 'free-market' capitalism destroys food self-sufficiency, with the land taken away from subsistence farmers to allow for intensive farming of export cash crops for profit. Inevitably the new, less labour intensive plantations have employed only a fraction of the displaced population. Restricted supply of manufactured goods by first world oligopolies (to keep prices high) has meant that capital export has employed only a fraction of the rest, despite below subsistence wages and appalling working conditions. Failure of land reform and low prices for industrially produced and subsidised food imports from the developed world have typically prevented such displaced populations from succeeding as small-scale market food producers because of the difficulties of competing with such cheap imports and local agribusiness.

Recent developments

In the last few years the historical decline in food prices has been reversed with prolonged supply shortages and price spikes exposing billions more people to hunger and malnutrition. In particular, oil price rises have contributed to massive increases in transport and fertiliser costs, with the cost of diammonium phosphate, a commonly used corn fertiliser, for example, rising from under US$300 per ton in 2007 to US$792 in February 2008 (Spencer, 2008). So too did oil price hikes contribute to increased demand for bio-fuels, diverting grain from food production and putting up food costs. As Maslin (2009, p. 97) points out, 'the amount of grain required to fill the tank of a car with ethanol could feed a person for a year'.

With the collapse of the US housing market and the subsequent crisis in the financial sector, speculation by investment funds shifted increasingly into commodities, seen as safer and less vulnerable to the developing recession. The deregulation of investment banking by 1991, with the abolition of rules limiting such banks' speculative position in commodities, paved the way for this development (Mason, 2009, p. 108).

Increasing demand for agricultural commodities, due to rising populations, and increased income and meat consumption in Asia, appealed to speculators. Anticipating ongoing price increases, they bought futures contracts (guaranteeing particular fixed prices) on crops months before harvest – so as to able to cash in on these and further price increases due to droughts, blights and other natural disasters. A weakening US dollar coupled with the relative strengthening of the currencies of importing countries contributed to significant increases in US grain exports, with reduced inventories at home, which further encouraged such speculation.

As Wilson noted in April 2008, investment in grain and livestock futures

> more than doubled to about $65 billion from $25 billion in November [2007] according to consultant AgResource Co in Chicago. The buying of crop futures alone is about half the combined value of the corn, soybeans and wheat grown in the US, the world's largest exporter of all three commodities.

Speculation itself pushed up prices and generated further speculation as others rushed to cash in. Commodity investors competed with governments and consumers for dwindling food supplies.

With 'the demand for futures exceeding the demand for cash grains', grain distributors and processors ran up debts to finance bigger inventories.

With prices of rice, corn and soybeans as well as wheat climbing to record levels, partly because of continued drought in Australia, a freeze in Kansas, poor crops in China and increased demand for livestock feed, as well as rampant speculation and hoarding, increased oil costs and demand for biofuels, 2008 saw food-related protests and riots in Haiti, Indonesia, West Africa, Morocco, Bangladesh, Mexico and Egypt.

In many cases the most desperate populations were victims of grain hoarding by local distributors, withholding stocks till they could be sold at the highest possible prices. In early 2008 both the Indian and South African Governments, threatened with popular revolt, attempted to crack down on such hoarding which was contributing to massive wheat price inflation.

By late 2008, the deepening recession drove hedge funds and institutional investors to pull billions of dollars out of commodities markets to repay loans, thereby contributing to falling prices (Mason, 2009, p. 108). Declining world demand has brought down oil prices and therefore also prices of agricultural chemicals produced with oil. Declining oil reserves will inevitably increase prices again in the future. Increasing economic and ecological instability coupled with rising demand will contribute to increasing volatility in grain prices, and increasing difficulties for poorer people struggling to survive.

Industrial agriculture

Walden Bello (2009, p. 22) describes how a world market in agricultural produce originally developed through the juxtaposition of 'two food grids under the institutional canopy of a global free trade system promoted by Great Britain'. On the one hand, large capital intensive farms came increasingly to replace small family farming, specialising in the production of wheat and livestock in the European settler colonies of the US, Canada, Australia and Argentina. Such foodstuffs were increasingly exported to provide 'the wage foods for the industrialising metropolitan economies of Europe'.

Meanwhile, in the colonial territories of the global South, a complementary system of plantation production provided tropical crops of 'sugar, tobacco, coffee, tea and cocoa', along with 'cotton, timber, rubber, indigo and jute' for export to 'expanding European markets'. Such plantations coexisted with small-scale peasant production of basic food crops, including rice, for the subsistence of both the family farmers themselves and the plantation workers. The expansion of the export economy forced such small producers to struggle to 'squeeze more and more from less and less land' (Bello, 2009, p. 23).

In the developed world, particularly in the US from the late 1800s, government intervention contributed to the development of new faster growing, higher yielding plant strains. Such types required higher levels of nutrition than could be provided by traditional methods of manuring and crop rotating with the use of biological nitrogen fixation, leading to soil destruction. This contributed to the development and application of the Haber-Bosch system of industrial nitrogen fixation to produce fertilisers, which, combined with increasing mechanisation and specialisation, led to big productivity gains in grain production after 1950, allowing fewer people to produce much more output (P. Roberts, 2008, p. 22).

Increased grain production, in turn, allowed for grain fed cattle, pigs and chickens to be 'processed' at much greater densities in 'concentrated animal feeding operations or CAFOs' (P. Roberts, 2008, pp. 23-4). The price of the hugely increased efficiency of a 'single integrated supply chain' from feed mills to slaughterhouse ('under the control of a single company') was increasing misery for escalating numbers of animals (P. Roberts, 2008, p. 24).

A continued rise in output has increased meat consumption by five times, which is more than twice as fast as population expansion, between 1950 and 2008. Consumption levels have risen across the developing world, including even the poorest countries, with China's consumption doubling in the last 20 years.

These developments have contributed to a growth in land consolidation by larger entrepreneurial farming operations able to effectively exploit the new technologies, and thereby accelerated the displacement of small-scale and subsistence farmers from the land. As Roberts points out, in 1885 more than half of the US population were farmers, but this had fallen to 3% by 1985, producing 40% of world corn output (P. Roberts, 2008, p. 23).

In the 1920s, governments in the US and Europe began to introduce agricultural price supports and production limits to try to cope with increasingly destructive overproduction (and price falls) – as well as periodic crop failures – in grain markets produced by increasingly high output farming techniques. A wave of farm collapses in the US persuaded US leaders that a free market in food would have unacceptable longer-term consequences for the country. In order to moderate prices by stabilising supply, the government began to pay farmers not to use all available arable land – so as to reduce supply – while also guaranteeing a minimum per bushel price for their grain.

> If market prices for grain fell below that target, governments would pay the farmer the target price and put the grain in a national reserve, to be released in time of shortage. (P. Roberts, 2008, pp. 117–18)

In practice, such guaranteed prices encouraged overproduction, with farmers paid for whatever they produced, despite falling prices. 'US grain volume rose faster than the government could pull acres out of production', with government stockpiles growing ever larger (P. Roberts, 2008, p. 118).

The global South

After the Second World War, the post colonial Bretton Woods regime included high tariffs to protect developing world industry and agriculture, with strict regulation of foreign investment. The US permitted such policies in exchange for anti-communist military alliances (Bello, 2009, p. 26) with some countries including Japan, Korea and Taiwan.

Agricultural exporters in the global south, typically larger operations, were encouraged to increase production to provide foreign currency for the import of producer goods, while smaller agricultural producers were heavily taxed to support import substituting industrial development (Bello, 2009, p. 24).

Peasant farmers did receive support from import bans, quotas and high tariffs, along with state subsidies, cheap credit, guaranteed prices and state insurance, and investment in rural water supply, health and educational facilities in some areas.

But many developing world governments welcomed cheap subsidised US food exports and food aid to support rapid industrialisation, which undermined local peasant production. Exporters benefited at the expense of smaller producers, who were forced into more labour intensive production in smaller fields (Bello, 2009, p. 24).

Fearing peasant support for left-wing governments, the US actually supported significant land reform, restoring land to the poor and landless, starting in strategically important Taiwan and Korea in the 1950s, and moving to Vietnam and Chile in the 1960s. The failure of such moves to prevent left-wingers coming to power in key regions encouraged the shift to other policies (Bello, 2009, p. 28).

In many parts of the developing world, concentrated land ownership served to motivate social revolution both directly, because of the poverty and slavery of the mass of the population denied land ownership, and indirectly, because of the failure of the large semi-feudal agricultural estates to feed those populations, with crops exported for profit. Large farms in Latin America, for example, tended to significantly underutilise the land and convert arable land to cattle farming for export (Weis, 2007, p. 96).

When mass peasant mobilisations, revolutions or democratic elections brought reforming governments to power, promising significant land reforms, the US used subversion and violence to remove (or try to remove) such governments and create ruthless police states to protect their agro-exports and mineral interests (e.g. in Bolivia, Brazil, Chile, Cuba, the

Dominican Republic, El Salvador, Granada, Guatemala, Jamaica, Nicaragua, Peru and Venezuela) (Weis, 2007, p. 97).

From the late 1960s, the US leadership became strong supporters of what came to be called the Green Revolution (GR) involving the insertion of new higher yielding seed varieties (of rice, wheat, maize and other crops) coupled with fertilisers, agrochemicals, irrigation and machinery into the agricultural economies of Asia, Africa and Latin America. This significant technology transfer was initially financed through substantial Western loans and aid contributions, justified both by the cold war and by future prospects of sales of US produced agricultural inputs to the countries concerned.

The GR can be seen as an alternative, non-military means of combating social revolution in the developing world. As well as providing increased revenues for Western – and particularly US – firms supplying industrial inputs, it also provided 'a means of increasing food production without upsetting entrenched interests' (Weis, 2007, p. 94). In particular, there was no significant land reform, breaking up the big estates of the wealthy into smaller, labour intensive production units for the poor. On the contrary, the new technologies encouraged and accelerated concentrated land ownership. At the same time, they contributed to the destruction of soil fertility, to pollution and rapid depletion of finite water resources.

Ecological considerations

Defenders of the current system argue that industrial agriculture has made a key contribution in boosting food output to keep pace with the increasing world population, which is moving towards seven billion. They argue that industrial agriculture is needed to continue increasing output in the future to cope with a predicted population of nine billion or more by 2050. The GR showed the way with new seed types (produced by cross-breeding and deliberate mutation), fertilisers, agrochemicals, irrigation and mechanisation yielding high and stable crop yields, starting in the developed temperate world and rapidly spreading to the developing tropical world. This boosted revenue for the Western firms producing such agro-technological inputs, and increased productivity for all those using such inputs, with irrigated lands providing only one fifth of total farmland but two fifths of total crop production.

It is frequently claimed that such changes were necessary to allow for increased food production to avert widespread famine in the 1960s, particularly in parts of Asia. It is also claimed that 'traditional' methods were quite incapable of keeping pace with rising population. Traditional 'swidden' agriculture – the felling of a patch of forest and use of the resulting land for a short period until the soil is exhausted, before moving to clear a new patch – was sustainable only so long as low population density allowed the forest time to regrow. By contrast, it is argued that modern industrial agriculture can be sustained forever, through industrial nitrogen fixation and more fertiliser.

Defenders acknowledge that the new techniques have contributed to increased land consolidation by larger entrepreneurial farming operations that are able to exploit effectively the new technologies, and thereby accelerated the displacement of small-scale and subsistence farmers from the land. But this has been more than offset by the productivity gains achieved, which have contributed to agricultural exports and increased gross domestic product (GDP). Efficiency gains are set to increase further with greater use of genetically modified (GM) crops and animals, including the development of pest, herbicide and drought resistant crops, and animals able to exploit currently marginal environments. Greater productivity will reduce the need for more land to be turned over to farming and could, indeed, free up current farm land for other uses (Weis, 2007, pp. 163–5).

Critics point out that industrial agriculture is destroying all environments, pumping toxins into the air and soil and water, and accelerating global warming. Mechanised tillage accelerates soil erosion, with 40% of agricultural land now seriously degraded worldwide. Nitrogen and phosphorus fertilisers are increasingly losing their effectiveness and greater inputs mean greater run-off into groundwater, rivers, lakes and sea, sustaining algal blooms and producing dead zones.

Capitalist farming has driven a radical simplification of the natural ecological order through the development of monocultures and the depletion of soil micro-organisms. The use of synthetic fertilisers and other agrochemicals, intensive irrigation, enhanced seeds and farm machinery, concentrated animal feeds, animal antibiotics and hormones, and long distance transport of agricultural products have all contributed to this simplification and to related ecological problems. All of these things involve massive use of fossil fuels which means that they contribute to global warming and will become increasingly costly with further depletion of oil reserves.

In the past, self-sufficient and sustainable farming techniques were built around a functional diversity of soils, crops, trees and animals necessary to maintain an ecological balance around key nutrient cycles. Dense multi-cropping, with regular rotation, use of green manures, fallowing land, agro-forestry, careful seed selection and integration of small animals all contributed to such sustainability.

Crop yield efficiency and output per agricultural worker – highlighted by GR proponents – are not good measures of overall productive efficiency. We need to consider also the net output per unit area and the long-term sustainability of agricultural practices through recycling, reduction of toxicity and increase in soil carbon. Small farms on similar quality farmland can produce higher net output per unit area over longer periods, through more intensive land use, with complementary plant and animal species.

Shiva draws upon evidence from many parts of India to show that 'small biodiverse farms' using organic methods, including farms that have completely banished the direct or indirect involvement of fossil fuels and artificial fertiliser, have higher output per acre than monocultures. With fields of up to seven, nine and 12 different crops, in symbiotic interaction, with use of oxen for ploughing and fertiliser, such farms have soils high in organic matter, necessary fungi, earthworms and available nutrients, and such

> biodiverse systems are more resistant to droughts and floods because they have a higher water holding capacity, making them more adaptable to the effects of climate change. (Shiva, 2008, p. 112)

Far from the GR 'saving' the poor of the developing world, it actually saved the wealthy landowners from expropriation and returning the land to peasant producers, and the agrochemical suppliers from increasing marginalisation.

It's true that with inadequate opportunity for forest regrowth, swidden agriculture rapidly destroys soil fertility. But as Tim Flannery (2009, p. 140) points out, mechanised tropical forestry and agricultural practices greatly amplify the destruction. Most significantly, '70% of Amazon deforestation is attributed by the UN [in 2006] to the expansion of cattle farming, rather than agriculture or logging'. This has led to tropical regions becoming full of 'rank grassland growing on impoverished soil'. Small-scale sustainable forest agriculture is perfectly possible.

Defenders of the current system endorse the further expansion of high meat diets. However, land can be saved by reducing meat consumption. More than two thirds of arable land is currently devoted to livestock production through pasturage and production of animal food crops. From 4.5 to 20 units of cereal food by weight are needed to produce one unit of meat. Reduced meat consumption would improve human health through reduced consumption of saturated fats, cholesterol, pesticides, nitrites, pharmaceuticals and hormones. It would reduce the likelihood of animal and food borne pathogens – including avian influenza and bovine spongiform encephalopathy (BSE) – affecting human populations. It would reduce desertification of marginal lands by animal hooves and it could save some fish stocks from extinction, given the big share of fish currently going to animal feed.

Factory farmed animals typically carry a heavy burden of fat, which as well as being a health problem in itself, also bio-accumulates organic pollutants (including dioxins and other organochlorides) and agrochemicals from industrial monocultures of feed crops and animals. GM food crops can transfer dangerous transgenes and viruses to factory farmed animals and on to humans. The widespread use of antibiotics to combat disease-producing living conditions for such animals has contributed to antibiotic resistance in bacteria, seriously threatening human health. A Pew Research Centre report referred to 'the continual cycling of viruses' in industrial farms, increasing opportunities for the development of 'more efficient human to human transmission' (http://www.guardian.co.uk/commentisfree/2009/apr/27/swine-flu-mexico-health).

Monocultures

Increasing use of fertilisers contributed to monoculture as it was found that different crops responded better to different ratios of nitrogen, phosphorus and potassium in the fertiliser. Monocultures, along with artificial NPK fertilisers, contributed to 'increased susceptibility of plants to diseases and insect attack' (Ashton and Laura, 1998, p. 31). Crops grown in soils low in humus suffer more from pests. Monocultures are basically 'plagues' of plants which produce corresponding 'plagues' of insects and plant diseases. So increased fertiliser use has led to increased use of toxic pesticides.

Greater use of pesticides leads to accelerated development of resistance, and use of more and more toxic chemicals to try to keep pace. It renders crops poisonous to 'good' insects as well as 'bad', to soil microbes, crucial to soil fertility and productivity, and to humans. Insecticides 'routinely follow fertilizers out of the soils and into water supplies', and the widely used organophosphates pass easily through human skin to cause heart irregularities, mental impairment and other serious problems, with 97 000 cases of exposure reported to US health agencies in 2002 (P. Roberts, 2008, p. 218). In 1990, around the world, three million people were recorded as suffering acute insecticide poisoning with 220 000 deaths attributable to exposure to agricultural chemicals.

A major issue is the role of big, mainly US and European corporations in controlling the production and distribution of foods worldwide; producing, refining, combining and marketing. This includes control of patents as well as production and long distance transport systems, and applies to seeds, fertilisers and other agrochemicals, livestock antibiotics (with US livestock consuming eight times more antibiotics than humans do) and farm machines. As Weis points out, in 2004, the top ten transnational agrochemical corporations (TNCs) controlled 84% of the US$35 billion global market led by Bayer, Syngenta, BASF, Dow, Monsanto and DuPont (Weis, 2007, p. 72).

This has meant a massive homogenisation of food production in the service of market efficiency and profitability, around what has been called the industrial grain/livestock complex. Large-scale cash cropping has gone along with a major decline in genetic diversity within and between farmed animal and plant species in the twentieth century, so that now 10 crops account for three quarters of all plant calories worldwide. The apparently great diversity of foodstuffs on offer in a modern supermarket is actually produced from a very limited range of basic components.

Similar homogenisation has been pursued in relation to farm animals, with 88% of animal flesh by volume now coming from pigs (39%), chickens (26%) and cattle (23%). At the same time, significant productivity increases have been achieved in relation to farm animals through the industrialisation of animal farming. This has included the warehousing of large populations (now 40% of meat production by volume), with controlled acceleration of growth and biorhythms. Broiler chickens are now ready to eat in weeks and pigs in six months. World chicken production has doubled since 1990. An increasing output of food plants has been devoted to animal feed, particularly maize and soy feeding factory animals.

Monocultures are not just vulnerable to insects, but also to disease and to droughts, storms and floods driven by climate change. With big business corporations selling concentrated packages of commodity inputs, farmers lose traditional knowledge and skills; they lose the capacity to maintain diversity and respond to changing ecological conditions. GM crops mean still greater genetic simplification – and consequent vulnerability – and threaten other species, human health and ecosystems (see Chapter 6).

Dryland salinity

There is a number of different but interdependent water-related problems affecting current and future agricultural production and public health. Of major significance here are issues of destruction of agricultural land and fertility through salination, alkalisation, waterlogging and land subsidence.

So-called dryland salinity is an issue particularly where European settler farmers have indiscriminately destroyed native vegetation to make way for imported grain monocultures.

Dryland salinity is generally caused by the removal of deep rooted native vegetation, including perennial trees, shrubs and grasses and their replacement by annual crops and pastures using less water.

> Dryland salinity has been a major problem in the Dakotas of North America and the western prairies of Canada. The salt lies underground from the weathering of ancient rocks. The perennial grasses which covered these regions were well adapted to absorb . . . available moisture . . . the imposition onto these landscapes of large-scale wheat farming upset the balance. The shallow rooted wheat crops allowed seepage of moisture underground, causing the water table to rise and bring salt to the surface. The resulting environmental damage [saw] over a million hectares lost to production [by 1991] [with] a growing death-rate among a range of wildlife. (Beresford et al., 2004, pp. 3–4)

In Australia, bad land management and clearing of native vegetation have contributed to large-scale 'soil erosion, soil acidity and declining soil structure' as well as salinity. Indeed, a greater percentage of native vegetation has been cleared in Australia than in any other developed country. In Western Australia, between the 1920s and early 1980s, more than 30 million hectares of deep rooted perennial, native vegetation and massive terrestrial and aquatic biodiversity were destroyed and replaced by shallow rooted cereal crops, particularly wheat, along with some pasture for sheep. The region became a major agricultural exporter.

The native vegetation had absorbed moisture before it reached the water table and evapo-transpired it back into the atmosphere, so had been well adapted to surviving flooding and drought. The annual crops and pastures, however, used much less of incoming rainfall with the unused water 'either running off or infiltrating beyond the root zone and accumulating in ground water' loaded with salt 'carried inland' from the sea over thousands of years (Beresford et al., 2004, p. 6). As saline groundwater rose close to the surface and entered plant root zones, it killed both remaining native plants and crops and pastures and led to increasing waterlogging. Salt was concentrated on the soil surface by evaporation, further damaging such soils and draining into streams, rivers and lakes, 'degrading wetland habitats and water resources' (Beresford et al., 2004, p. 7).

With State and federal governments failing to act or even to acknowledge the problem for a long period, and such failure latterly supported by neoliberal 'non-interventionist' ideologies and policies, landholders responded 'by increasing production on unaffected lands'. As Short and McConnell note

> they did this with improved technology, farm rationalisation, higher yielding varieties and other agricultural system advances or simply clearing more land and thus perpetuating the cycle. This will be increasingly difficult in the future as the resource base continues to deteriorate. (Short and McConnell, 2000, pp. 21–53)

In Western Australia (WA), two million hectares were affected by 1996 and another four million hectares under imminent threat with one hectare lost to salinity every hour at the beginning of the twenty-first century, with groundwater continuing to rise 'at an average rate of 20 cm per year in the south west agricultural region as it is recharged by rainfall' (Beresford et al., 2004, p. 7). *The National Land and Water Resources Audit* found 5.7 million hectares Australia-wide already affected or likely to be affected by dryland salinity in the near future, with over 17 million hectares affected by 2050 without radical action (http://www.nlwra.gov.au/national-land-and-water-resources-audit/land-salinity).

This case illustrates some general principles of twentieth century capitalist agricultural development. On the one hand, the misleading nature of food prices determined by the cost of market valued inputs into the production process – land, labour and capital goods. As Pannell, Lefroy and McFarlane point out, in 2000, it would have cost WA farmers up to $1 million per farm to finance the revegetation needed to begin to restore the fertility of the land (2004). On the other hand, it shows how technological advance, which is not based upon solid ecological foundations, undermines sustainability in the longer term, and ultimately contributes to a deepening world food crisis.

Irrigation

Declining water supplies, due to inefficiency and wastage, depletion of resources and climate change pose a major and immediate threat to world food production, especially in light of increased population and demand (P. Roberts, 2008, p. xx). Many see this as a major source of global conflict in the near future.

Irrigation has been a particular problem area, with massive overuse of limited groundwater supplies and wastage of water through evaporation. As with dryland salinity, badly designed or operated irrigation systems lead to rising groundwater levels, bringing salt to the surface as irrigation water passes through the root zone of plants to add to the water table. In 2006, the Food and Agriculture Organization of the UN estimated 10% of irrigated land worldwide was severely damaged by waterlogging and salination (Weis, 2007, p. 32).

Industrial agriculture is responsible for the major problems as it has driven increasing dependence upon high levels of irrigation. Agricultural research has contributed to rising productivity in monocultures of rice, wheat and maize, with world output increasing three times from 1950–90. These three crops now account for half of the calories consumed worldwide. But this has depended upon a nearly three times greater use of irrigation water, as well as ten times greater use of fertilisers.

In India, GR rice needs eight to 12 times more water than similar crops of the more nutritious millet. 'India could grow 4 times the food it does if it were to cultivate millet more widely' (Shiva, 2008, p. 110).

'Soil microorganisms maintain soil structure, contribute to the biodegradation of dead plants and animals and fix nitrogen. They are the key to soil fertility' (Shiva, 2008, p. 101). They are destroyed by heavy use of synthetic fertilisers. 'Fertilizers block the soil capillaries which supply nutrients and water to plants.' 'Infiltration of rain is stopped, runoff increases and soil faces droughts.' So fertilised crops require ever 'more irrigation and ever more fossil fuels for pumping groundwater' (Shiva, 2008, p. 103).

Excess nitrogen in the root zone also prevents nutrients getting into the plants. This has led to serious deficiencies in iron, zinc, vitamin A and other micronutrients where the new technologies have displaced local foods (Ashton and Laura, 1998, p. 33).

The extension of the GR has depended upon ever greater water usage, with agriculture now responsible for nearly three quarters of all global fresh water use, and 1000 tons of water needed on average to produce one ton of grain.

As Mehrotra and Delamonica point out, the 'success' of the GR has depended upon massive government subsidies of water costs to wealthy landlords, able to exert significant political influence. This has, in turn, contributed to massive wastage of such cheap water due to defective irrigation systems.

> Many of the irrigation canals in low income countries are crude ditches dug out of the fields. Much of the water never reaches the targeted crops. Worse still, it is absorbed into ditches and later leads to salinization which destroys the fertility of the soil, or it lies in stagnant pools that provide breeding grounds for insects . . . if the ditches are lined with low-cost tile, the water can be directed efficiently to the targeted fields . . . (2007, p. 33)

In Australia, extensive use of low cost irrigation for large-scale cotton and rice farming, particularly across 25 million hectares of the Murray River basin in Queensland and New South Wales (NSW), has accelerated salination, with more and more salt getting into the river in drainage water undermining drinking water supplies in Adelaide and elsewhere. Huge amounts of water have been lost through evaporation from such irrigation projects, and such water diversion upstream kills the wetlands downstream (Pearce, 2006, p. 251).

Dams and flooding and river diversion for irrigation have pushed 30–60 million people off their lands in the last century and destroyed ecosystems and biodiversity around the world. Such irrigation has contributed to waterborne diseases.

Dense livestock also consume and pollute masses of water, with 3000 litres of water (enough for an average household for one month) used to produce one kilogram of US beef. Factory pigs consume 132 litres per day and a typical US slaughterhouse uses the same amount of water in a day as 25 000 householders. One thousand litres of water is used per large animal killed, with masses of waste water full of blood, faeces, fat, grease and cleaning solvents released into the environment. One hundred and thirty times more animal manure is produced in the US than human sewage by volume, and such manure is the major polluter of rivers and streams (Weis, 2007, pp. 33–4).

The rural poor displaced by the GR don't all go to the cities. In desperation they push agriculture further into marginal, hilly, erosion-prone arid land where soil fertility is difficult to maintain. They destroy rainforests, exposing nutrient poor soils that are rapidly turned to desert, and thereby contribute to large-scale flooding and mud slides.

The Green Revolution in the Punjab

As Vandana Shiva (1991, pp. 1–2) observes, the Indian Punjab 'is frequently cited as the GR's most celebrated success story'. Her research in the late 1980s found that 'two decades of the GR' had 'left the Punjab riddled with discontent and violence', with 'diseased soils, pest-infested crops, waterlogged deserts and indebted and discontented farmers'.

She found that the rate of growth of total crop production in India was higher prior to the GR, when the national government was encouraging ecologically responsible and sustainable development on small peasant farms. As elsewhere, short-term productivity gains, with GR wheat for example initially yielding 40% more than traditional varieties, along with government subsidies, favoured the first to take up the new technologies. Such gains were not maintained as 'a few years of bumper harvests' were followed by large-scale crop failure due to insect attack, micronutrient deficiencies and soil degradation (Shiva, 2008, p. 102).

Initial gains in output were insignificant compared to the huge increases in costly inputs. Measurement of output neglected the huge losses of 'nonmarketable' elements of crops, including fodder for animals and organic fertilisers for soils, and the very high water content of the new varieties (Shiva, 1991, p. 2). Increased susceptibility to pests led to ever increasing use of dangerous insecticides and fungicides. Soil destruction created greater and greater areas of 'wasteland' (Shiva, 1991, p. 5). Increased demand for diminishing water resources – to feed crops needing three or more times more water than traditional varieties – encouraged 'central control over large scale storage systems... leading to... local and inter-state water conflicts' (Shiva, 1991, p. 6). 'The GR has also resulted in soil toxicity by introducing excess quantities of... fluorine, aluminium, boron, molybdenum and selenium... into the ecosystem' (Shiva, 2008, p. 102).

Initially, 'high subsidies and price support' favoured the bigger farms, most actively embracing the new technologies. Later reduction in such subsidies led to declining real incomes and 'increasing indebtedness' for farmers (Shiva, 1991, p. 5).

As Pearce (2006, p. 58, p. 59) points out, while other Asian countries have emptied rivers into irrigation canals, India's GR has depended upon tapping underground reserves of water, with US$12 billion spent on pumps and boreholes over the past 20 years to irrigate at least two thirds of all of India's crops, including water-heavy crops such as rice, sugar cane, alfalfa and cotton.

One estimate is 100 cubic kilometres more used every year than is replaced by rain. Increasingly the aquifers are running dry with '200 million people facing a water-less future'. Thousands of farmers have committed suicide, with millions more leaving the land to crowd into urban slums. As water levels fall, there is increasing demand for electricity to pump the remaining deeper water to the surface.

Sustainability

All of the problems of industrial agriculture are interrelated, and all are intimately bound up with issues of climate change (explored more fully in Chapter 7). As Shiva notes, when the Stern Review identified land use as responsible for 18% of greenhouse gas emissions worldwide, agriculture for 14% and transport for 14%, it failed to identify the central role of industrial agriculture and corporate globalisation in all of these areas.

Much of the land use figure comes from the destruction of tropical forests to make way for cash crops and livestock. Millions of hectares of rainforest are being burned to grow soy as export cattle food, with armed gangs using slave labour to cultivate it (Shiva, 2008, p. 122). Much of the transport figure comes from increasing movement of foods around the world. A Danish Government study found that 10 kg of CO_2 were produced (on average) by moving each kg of food (from farm to consumer). Much of the agricultural figure comes from use of fossil fuels to make fertilisers and run farm machinery, and 'conversion of excess nitrogen fertilizers to nitrous oxide' in industrial agriculture, along with direct emission of CO_2, methane and nitrous oxide from soils, manures and livestock, from the decay of organic material in soils and from the burning of waste biomass' (Maslin, 2009, p. 105). Nitrous oxide is a much more potent greenhouse gas than carbon dioxide.

As oil prices have risen the push for more bio-fuels and biodegradable plastics has accelerated deforestation in the tropics, further increasing global warming, and nations and corporations have planted more palms and soya to cash in on the new markets. Higher temperatures produced by global warming will contribute to increased aridity in areas already subject to water stress.

Rising sea levels due to global warming will increasingly inundate major farming regions, including Bangladesh and the Nile delta. In Bangladesh, in particular, land subsidence (of 2.5 cm per year) due to a sixfold increase in fresh water extraction for agriculture in the 1980s will accelerate such inundation (Maslin, 2009, p. 84). Coastal aquifers will be polluted with salt, and increasing areas of farming land will be lost. At the same time, hotter seas will produce ever more and greater tornadoes, tropical storms, hurricanes, typhoons and floods, devastating crops, land and infrastructure.

With the soil's huge capacity for carbon storage (and increased carbon boosting soil fertility, resilience and retention of moisture and nutrients), agriculture should be contributing to reduced atmospheric CO_2 levels, through conservation tillage – or no till – agro-forestry, bio-oils and composts replacing fertilisers etc. Uptake of sustainable technologies has been slow in many regions and deforestation continues, to provide more land for farming.

The role of the IMF and the World Bank

In the early 1970s guaranteed price policies were abandoned in the US in an attempt to bring down agricultural costs, by increasing competition driven concentration and rationalisation, to be able to compete effectively in expanding world grain markets particularly in Asia and Latin America. But with massive overproduction and prices falling below costs, the government stepped in to 'pay farmers the difference between their production costs and world market prices by way of a per bushel "deficiency" payment' (P. Roberts, 2008, p. 121).

This led to further accelerated overproduction, with subsidised US farmers putting ever more grain into world markets, bringing world prices down, as farmers used subsidies to 'buy more fertilizers and pesticides and better seeds to grow even more bushels' (P. Roberts, 2008, p. 121). With subsidies failing to keep up with rising costs of land and inputs, and continued rejection of genetically modified US crops in overseas markets, leading to unsold corn and further price falls, smaller farms have fallen by the wayside.

The US saw the emerging economies of Asia and South America as huge potential markets for surplus US grain. However, China and India used their political and economic power to keep US exports out and pursue agricultural self-sufficiency. Other countries 'shielded their farmers behind barriers in the form of heavy taxes or tariffs on imported grain', and were increasingly vigorously criticised by the US for doing so, particularly with the rise to dominance of neoliberal free trade ideology in the 1980s.

Through its control of the International Monetary Fund (IMF) and the World Bank, the US leadership took advantage of the debt crisis, and the subsequent 'Asian meltdown' (produced by its own policies of enforced financial liberalisation and deregulation) to impose yet more liberalisation and deregulation with debtor nations 'required to restructure their economies along free-market lines' (P. Roberts, 2008, p. 128). What were described as 'trade-distorting' farm subsidies were to be removed – supposedly leading to modernisation and rationalisation of production. Together with enforced currency devaluations this was supposed to boost agricultural exports. At the same time, 'they [debtor nations] were also required to open their own markets to more imports... of fertilizer and other inputs for their new high volume farm sector, [and] also more grain' (P. Roberts, 2008, p. 129) from 'low-cost' US and European producers. No matter that such 'low cost' was itself in part the product of massive subsidies.

Where previous land reform, government support and high tariffs had still provided some protection for smaller farming operations, the IMF and the World Bank enforced structural adjustment in the 1980s and entry into free trade agreements in the 1990s further radically undermined the position of such farmers. In Mexico, for example, government expropriation of some larger holdings to provide communal land for landless farmers, creation of non-alienable collective property and substantial state support for agriculture in the form of low cost credit, floor prices, subsidies and rural infrastructure, had earlier slowed the spread of capitalist (export) farming, despite the GR.

Between 1981 and 1986 such government spending was cut by half as billions flowed overseas in debt servicing, with dismantling of support for poorer farmers. With Mexico's entry into the North American Free Trade Agreement (NAFTA) in 1994, more tariff reductions further reduced government funds available for rural development while allowing masses of cheap grain imports – particularly subsidised US grain – to cut local prices and undermine local production, driving poorer farmers into the cities and across the border into the US, and increasing dependence upon imported foodstuffs.

By 2007, similar developments around the world meant massively increased vulnerability of poorer populations to big food price rises in international markets, due to declining yields, increased oil prices and bio-fuel production, increased demand, monopoly power of big global food corporations and massive speculation. As Shiva points out, now that 'global corporations... have created import dependency, they are increasing prices... while millions go hungry, corporate profits have increased' with a 30% increase for Cargill and 44% for Monsanto in 2007 (Shiva, 2008, p. 96).

As she also notes, 'energy production for corn production in the US is 176 times more per hectare than on a traditional farm in Mexico and 33 times more per kilo' (Shiva, 2008, p. 99). With fossil fuels providing the bulk of this energy, the implications for the long-term sustainability of such a shift away from small-scale peasant production are clear.

The death of the peasantry

In the developed world a substantial amount of output now comes from very large-scale mechanised grain–livestock producers with a per farmer output around 2000 times that of small-scale developing world farmers who have no access to new technology, and are typically engaged in purely manual labour on poor land without any government support. It is these bigger operations that generally receive the bulk of government subsidies – over US$200 billion per annum in 2006 – approaching a billion dollars per day versus US$1 billion dollars per year in agricultural aid to the developing world.

Small farmers in the developing world struggle to compete with larger-scale farms benefiting from industrial methods, in their own countries, as well as with overseas producers. Litvinoff and Madeley cite the case of bananas

> The world banana trade – worth more than US$5 billion a year – is dominated by 5 transnationals, Dole, Del Monte, Chiquita, Fyffes and Noboa. They have driven costs down by sourcing supplies from large Latin American and West African plantations that pay rock bottom wages, where working conditions are abusive and the fruit is drenched dangerously with pesticides. The smaller independent banana growers – such as hillside farmers in the Caribbean – have been unable to compete on price and have lost market share. (2007, p. 12)

The destruction of subsistence farming to make way for large-scale, industrial export cash cropping, along with the destruction of smaller farms unable to compete with larger operations and subsidised imports, has driven the 'death of the peasantry' at an accelerated rate, creating a vast latent reserve army of desperate and destitute people in the less developed world. Between 1950 and 1990 agricultural employment in the developing world fell from 80 to 60% of the workforce, with urban populations increasing by 36% in the 1990s.

The destruction of agricultural subsistence has contributed to increasing dependence of poor people upon imports of basic foodstuffs particularly rice and flour. Food import dependence is greatest in Asia and Africa, with the fastest population growth, the greatest number of farmers and the greatest poverty. Relying on raw material exports with declining terms of trade to finance the import of necessities of day-to-day survival is not a good situation. As we have seen recently, food price increases due to increased cost of oil inputs have brought famine in many of these regions.

Small farm households still account for two fifths of world population. The continued expansion of 'free markets' in agricultural products, imposed by the World Trade Organization (WTO), and the pressures towards cash cropping to service developing world debt, will lead to the elimination of billions more poor farmers in future decades. There are few urban jobs for them now, with no likelihood of a big increase in jobs in the future and no provision for adequate (or typically any) social security. At the same time, this desperate pool of unemployed people keeps the power of labour and the cost of wages down on a global scale, as they are 'available' for ultra low cost urban employment in jobs transferred from other areas.

Such displaced and desperate people are denied opportunities for migration to other parts of the world. With the population racing ahead of employment, housing and infrastructure, these people will add to current urban slum populations, without adequate housing or sanitation, safe water or food. There are already a billion such people, and the figure is likely to reach two billion by 2030, according to the UN.

More free trade?

It is frequently argued that things would improve significantly for the poor of the developing world if there was 'genuine' global 'free trade' in food products, with developing world producers allowed unrestricted access to developed world markets. Both the then head of the WTO, Michael Moore, and the then World Bank president, James Wolfenson, referred in 2002 to the 'heavy burden' imposed on the developing world citizenry by agricultural subsidies in the developed world, which prevented any such 'level playing field' and undermined the benefits of development assistance (Weis, 2007, p. 146).

In fact, as Weis points out, in the context of current patterns of land ownership and political power, but without radical political change and land reform, relaxation of developed world subsidies and tariffs could function mainly to increase the wealth and power of landholding elites at the expense of still more poorer people without land or struggling to survive on small farms. While the big agro TNCs have been 'major beneficiaries' of subsidy regimes, 'they are most concerned with market access and would happily trade a measure of subsidy discipline for liberalization', with their ability 'to source cheaply...enhanced when the world's farmers are locked into borderless competition with one another, without bounds on the volume of farm production' (Weis, 2007, p. 158).

Litvinoff and Madeley argue the case for the positive potential of fair trade, with the distributors paying higher prices to poorer producers. Such prices

> cover production costs, plus provision for household members to enjoy a decent living standard and the cost of farm improvements and compliance with fairtrade standards, including the cost of belonging to a farm co-operative. (Litvinoff and Madeley, 2007, p. 12)

This means payment of a 'social premium' above market price when world prices are low. 'In recent years the floor price paid to producers for fairtrade-certified Arabica coffee has averaged $1.21 per pound, compared with an average world price of 70 cents a pound' (Litvinoff and Madeley, 2007, p. 12).

There are now more than a million 'small-scale producers and workers in 580 certified producer groups in 58 countries... actively involved in the system' (Litvinoff and Madeley, 2007, p. 43). This remains a very small number compared with the hundreds of millions living on less than US$2 a day, struggling as near slaves on export crop plantations, or without any employment.

Weis follows Bello in arguing that rather than debates about pushing for 'freer trade' within the WTO or through bilateral trade deals, the need is for

> real alternative arrangements, such as creating regional economic blocks or restructuring existing ones such as MERCOSUR or ASEAN to serve as effective engines of co-ordinated progress via policies that effectively subordinate trade to development. (Weis, 2007, p. 153)

China

Not all developing countries fell prey to US 'free market domination' enforced by structural adjustment and free trade agreements. The Chinese leadership, in particular, was able to use its political power and the huge internal productive capacity of the country to keep the Americans out and build agricultural self-sufficiency rather than grow luxury crops to pay for imported foodstuffs. However, they have recently changed direction in this area.

While the struggles of the peasantry brought the communist party to power, the party then took all agrarian surplus beyond basic survival needs to support the towns, and forced all the peasants into communes from 1958 to 1961. After 30 million died from malnutrition and starvation as a result of these developments, peasants were first allowed to lease land back from the communes and then, during the Cultural Revolution from 1966, taxation was reduced and land redistributed to individual households on an expanding scale.

The reforms of 1976 consolidated this privatisation process, with households allowed to retain what was left of their output for consumption or sale after selling a fixed proportion of their crop at a fixed price to the government or paying a money tax (Bello, 2009, p. 93).

With its borders closed to developed world food exports in the 1970s and 1980s, state power allowed a free internal market in agricultural produce to replace the earlier government monopsony. Under the new system, farmers sold some of their output to local customers and were encouraged to reinvest the profits in new seeds (adapted from Western varieties) and chemical fertilisers 'which the Chinese now use three times as heavily per acre as US farmers do' (P. Roberts, 2008, p. 125).

Labour, available in abundance, substituted for machinery, with China's 200 million farms remaining very small by Western standards. Instead of high levels of specialisation, most farmers plant several crops a year, and also keep livestock or operate fish farms. On this basis, as Roberts points out, China's 200 million family farms came to 'produce 20% more output than do the United States' 2 million farmers, on a land base less than three quarters as big as America's' (P. Roberts, 2008, p. 125). China also came to compete with the US in exporting food to other parts of Asia.

The period from 1978–84 was one of huge rural poverty reduction and increased output. After that, the state pushed up rural taxation and reduced rural support in a big push for accelerated industrialisation. Entry into the WTO in 2001 further worsened the position of farmers, with many studies suggesting massive loss of work and income for hundreds of millions of farmers, with farm households' position relative to urban households made much worse as a result of cheap imports (Bello, 2009, p. 93).

Heavy use of fertilisers contributes to global warming, soil degradation and water dependence and leaves the country vulnerable to declining oil stocks and climate change. Agriculture competes with urban development for limited water supplies and land, and water shortages and land loss have cut production on the North China Plain (Bello, 2009, p. 88).

Water consumption for agriculture in Eastern China

> now exceeds the sustainable flow by more than six hundred million tons a year, according to a 2001 report by the World Bank... According to the World Bank, even if China adopts a rigorous water-management system – with higher price incentives, better system efficiency, recycling of waste water and massive water transfers [as much as 270 million tons a year] from the wetter southern part of the country to the drier north, [as is planned] – the 3-H region will still face a water deficit of 600 million tons, or about 2 thirds of the total yearly flow of the Huang river. (Bello, 2009, pp. 229, 230)

With the population being allowed to rise to support increasing numbers of retirees, and increasing demand for meat amongst more affluent city dwellers, China now imports increasing quantities of soybeans as animal feed, while exporting its scarce water in the form of low cost grains and meat.

Severe winter storms in China pushed food prices 18% higher in early 2008 than a year earlier, creating serious hardship for millions of poor rural households. Faced with food riots, the government restricted exports and reduced tariffs on imported foodstuffs. This, in turn, contributed to rising world grain prices as considered earlier.

Cuba

China provides a model of labour intensive productivity. Current agricultural practices are not sustainable in the longer term, nor are they combating global warming. Cuba, on the other hand, has taken major steps towards genuinely sustainable organic farming, following the US economic blockade since the 1960s, and the food crisis it suffered through the collapse of the Soviet Union.

Restricted supplies of agrochemicals due to the blockade, motivated a shift of one third of Cuba's 11 million hectares of agricultural land to fully organic farming, with a third still using such chemicals and the remaining third half and half. As Mae-Wan Ho points out, 'the yields per hectare of the fully organic [were] equal to the fully agrochemical', providing 'clear evidence that organic agriculture can work on a large scale' (Ho, 1999, p. 144).

With the collapse of the Soviet Union, the supply of agribusiness inputs and grain traded by the USSR for Cuban fruit and sugar dried up. The regime responded by breaking up the big state run farms into smaller cooperatives, run on a sustainable, organic foundation. Large numbers of workers

> were reallocated from urban jobs to farms...thousands of produce gardens were planted...in cities. At universities and research centres scientists [found] ways to replace heavy industrial farm inputs. Oxen breeding programmes were expanded to replace tractors... (P. Roberts, 2008, p. 305)

Cuba now has approx 365 000 oxen, replacing 40 000 tractors (Shiva, 2008, p. 76).

> Cubans adapted...methods of integrated agro-ecological farming, including mixed livestock crop operations, crop rotations, inter-planting, and integrated pest management...to replace synthetic fertilizers and pesticides. [As a result of these developments] the country now leads most developing nations in nearly all nutrition and food-security categories. (P. Roberts, 2008, pp. 305–6)

Obesity issues

Major fast food corporations including Unilever, PepsiCo, Cargill, Nestlé, Altria, Sara Lee, Coca-Cola, Wendy's and Tyson market foods high in sweeteners and fat as well as salt and chemical additives, which are 'not only calorie-dense in themselves', but have been found to 'actually stimulate us to eat more'. Poorer people are targeted with massive TV advertising – particularly aimed at children – and more fast food outlets in poorer neighbourhoods selling cheaper foods containing more calories per kilo than more costly foods.

The actual quantity of food eaten per person in the US has been steadily increasing to the point where

> according to the US Centres of Disease Control, complications from obesity and related problems...cause 112,000 premature deaths and account for US$75 billion in extra medical costs in the US each year. (P. Roberts, 2008, p. 83)

Things are predicted to get worse with 'children growing obese in greater numbers and at earlier ages than before' (P. Roberts, 2008, p. 83). On a world scale, 'obesity afflicts a billion people – roughly the same number who are underfed' (P. Roberts, 2008, p. 84).

It seems that because humans evolved in a situation of relative food scarcity, the body's natural control systems tend to 'err on the side of encouraging overconsumption', (P. Roberts, 2008, p. 87) to make the best of brief periods of abundance. For most people this wasn't a problem in ages gone by because of continued scarcity. A deluge of ever cheaper and more accessible food in the developed world – and the US in particular from the 1890s to the 1960s and, even more so, from the 1980s – has radically changed this. By making food cheaper and easier to use the food industry has 'removed two of the biggest natural restraints on overconsumption' (P. Roberts, 2008, p. 87).

As considered in Chapter 3, the period from the 1980s was also the time when neoliberal policies functioned to create increasing unemployment and underemployment, and increased intensity of work. There was increased insecurity and fear of job loss, with reduced rights and powers and stagnant or declining real wages for working people. Greater inequality, disempowerment and insecurity contribute to excess consumption of 'comfort foods' meaning the kinds of sweet and fatty foods produced by the food processing industry.

The food industry has responded to increasing bad publicity and lawsuits by 'rolling out numerous reduced fat/low calorie products... since 2002 – and launching campaigns to improve consumer nutrition awareness' (P. Roberts, 2008, p. 98). As Roberts points out, 'such efforts are hard to take seriously' since 'the 4000 healthier products the industry touts represent just 7% of the products introduced between 2002 and 2006... [and] the food companies cannot afford to have us eat any less than we already do'. If Americans cut their daily intake by the 100 calories some leading nutritionists see as enough to turn around the obesity epidemic 'it would cost the industry between US$31 billion and US$36 billion in US sales alone' (P. Roberts, 2008, pp. 98–9).

Obese people are more prone to 'sleep disorders, blood clots, leg ulcers, pancreatic inflammation and hernias'.

> Heavier people put more stress on bones and joints... and extra padding... prevents the lungs from fully expending, leading to lower blood levels of oxygen and shortness of breath. Obesity also makes medical treatment harder... [It is linked to] higher rates of heart disease, both because the heart must work harder and because it contributes to higher levels of triglycerides and CDL with lower levels of HDLs and adult onset diabetes... with insulin's regulatory role [in maintaining blood glucose levels] diminished by high blood levels of fatty acids. (P. Roberts, 2008, p. 90)

An increase in obesity is implicated in an expanding epidemic of cancer. In 2009, the World Cancer Research Fund identified obesity – along with diets high in salt and processed meats – as a major cancer risk (http://www.newscientist.com/article/mg20126983.600-lifestyle-changes-could-cut-cancers-by-a-third.html).

Other food health issues

We have already touched upon some of the health consequences of heavy insecticide, herbicide and fertiliser use by industrial farming operations. The Food and Drug Administration (FDA) analysis of commonly consumed foods in the US continues to show high levels of contamination by a range of highly toxic and mutagenic organochloride and organophosphate pesticides, including substances like aldrin, dieldrin, hexachlorobenzene, lindane and others, banned from use years ago. Independent investigators, including the Washington DC based non-profit Environment Working Group, have found twice the levels of contamination reported by the FDA. In the mid 1990s they found 5.6 % of all fruits and vegetables contaminated with illegal levels of pesticides. As Burdon points out, regular UK Government tests show 'up to 3% of fruit and vegetables have pesticide levels above the legal limit' (Burdon, 2003, p. 75).

These compounds persist and accumulate in human fat and are released into the blood. When the World Wide Fund tested the blood of three generations of women from across Europe for a range of poisons and mutagens, they found that while the grandmothers carried the biggest burden of toxins – 63 poisons on average – including particularly pesticides, the next highest level was found amongst the granddaughters – 59 different poisons. While they had lower levels of some of the older, highly poisonous and long banned agricultural chemicals, nonetheless they still carried such chemicals – including DDT and lindane – as well as newer products such as perfluorinated chemicals from non-stick cookware (and neurotoxic PBDE – polybrominated diphenyl ethers – flame retardants) (Schapiro, 2007, pp. 126–7).

Cadmium is an extremely poisonous and carcinogenic heavy metal released into the air from burning wastes, from steel production and car exhausts. It ends up in the soil. It is also found in sewage sludge containing industrial waste. 'The cadmium level of phosphate fertilizers can commonly be as high as 100 ppm' (Ashton and Laura, 1998, p. 169). As Ashton and Laura point out, wheat and rice then concentrate cadmium from the soil. With widespread use of high cadmium fertiliser, a range of foods and drinks including alcohols, potatoes, carrots, beef and lamb have been found to have cadmium levels exceeding the maximum permitted concentration (MPC). Cadmium damages the kidneys and produces diabetes like symptoms. It is also strongly linked to prostate cancer (Ashton and Laura, 1998, p. 169).

Heavy use of nitrate fertiliser increases the nitrate content of fruit and vegetables. It also produces nitrate pollution of drinking water. Nitrates are converted to nitrites in the body which react with amines and amino acids as digestion products in the stomach to produce mutagenic and carcinogenic nitrosamines, leading to possible stomach, oesophageal or other cancers. Nitrite is used directly as a meat preservative and converted to nitrosamine during cooking (Burdon, 2003, pp. 82–3). Such preserved meats have been identified by the World Cancer Research Fund as a significant cause of human cancers.

Animal rights issues

Industrial farming practices subject animals to extreme physical and emotional suffering. In factory farms they are kept in small individual enclosures on metal grated or concrete floors or in very crowded shared spaces, deprived of fresh air and sunlight, fed concentrated foodstuffs, drugs and hormones. Factory hens, in particular, are crammed into spaces too small for them to spread their wings. Factory farmed animals are mutilated in various ways including de-beaking. Females are repeatedly inseminated and separated quickly from their young. Animals are transported long distances to slaughter without food or temperature control. They are pushed and prodded towards their deaths amidst the sounds and sights and smells of slaughter.

Defenders of factory farming point out that chicken, as a tasty, healthy and nutritious food, was once a luxury for ordinary people. These people could afford to eat it, if at all, only rarely. Now they can eat it regularly, due to the cost reductions brought about by intensive chicken farming. Organic and free range chickens remain significantly more expensive, and out of the reach of poorer people. As shed chickens are indoors, they are unlikely to come into conflict with the avian flu virus from overseas birds.

Critics point out that most supermarket chickens in Australia and the US are raised at high densities, with 30 000 or more chickens crammed into sheds 490 by 45 feet. Victorian legislation allows 40 kilos of meat per square meter which gives each chicken a space of less than one cubic foot. Mature birds are unable to move without pushing against other birds. They can't move their wings or escape from aggressive neighbours. Such environments are specifically tailored to provide the least floor space for the greatest return on investment.

High ammonia levels from their droppings lead to chronic respiratory diseases. Modern chickens grow three times as fast as earlier generations on a third as much food. As muscles and fat grow faster than bones, 26% of shed hens develop bone disease, leading to breakages and paralysis. Paralysed chickens then die of thirst and starvation. Overweight and without exercise, such chickens are also prone to heart attacks and other illnesses.

Breeding chickens are kept continually hungry since they have to live longer than the non-breeders and a similar diet would kill them too quickly. Most birds live only six weeks before they are sent to the slaughterhouse. Critics claim that they are generally pulled out of their boxes by one leg to go into crates for journeys that can last for hours. In the process limbs are often dislocated or broken. At the slaughterhouse they are stunned in electrified water baths but are often still conscious as their throats are cut or as they are put into boiling water.

Chicken sheds are heavy polluters, pumping out ammonia, with chicken waste attracting flies and getting washed into rivers and groundwater. This increases toxic nitrate levels in drinking water and promotes algal blooms in rivers, using up oxygen, killing fish and other river life. Jobs in the sheds are dangerous and poorly paid, and cramming chickens together in such conditions could further the development of avian flu viruses which could devastate human populations (Singer and Mason, 2006, Chapter 2).

Water

As Mehrotra and Delamonica point out

> over a sixth of humanity (1.1 billion people) are without access to safe water, and over a third (2.4 billion) lack adequate sanitation. Half the world's hospital beds are occupied by patients with water-borne diseases, implying that expensive curative services are being used to treat diseases that could easily have been prevented. (2007, p. 163)

Although in the 1990s more than 400 million city dwellers in the developing world gained access to water supply, the new supply failed to keep up with increasing urban populations, with an overall increase in those lacking any such access by 2000 (by 62 million). A comparable increase in water supply to rural populations led to an overall reduction in those lacking safe water during the same period (by 150 million leaving 915 million without clean water by 2000) (Mehrotra and Delamonica, 2007, p. 163). An additional 542 million urban dwellers and 252 million rural people gained access to basic sanitation in the 1990s. However, 2.4 billion city dwellers and 2 billion country people remained without adequate sanitation after 2000.

As noted earlier, irrigation for agriculture currently uses two thirds of fresh water supplies and most of the rest goes to industry. This leaves comparatively little for domestic use. Along with salinity issues and declining supplies, this, again, highlights the vital need for much less, and much more efficient, irrigation, to sustain production and free up water for other uses (Mehrotra and Delamonica, 2007, p. 163).

Chapter 2 considered the privatisation of water resources pushed by the IMF and World Bank. This meant that increasing numbers of poor people had to pay for their water directly, for the first time, or had to pay more, or were denied access to previously available water, leading to large-scale suffering, anger and increasing unrest. Mass protests have succeeded in reversing such privatisation in a number of areas. However, there is a long way to go to move towards rational, sustainable and just distribution of water resources around the world.

DISCUSSION TOPICS

1   On the basis of the evidence provided in this chapter, what do you see as the major issues and problems for future food security?
2   What ethical issues are raised in this chapter concerning eating habits in the industrialised West?

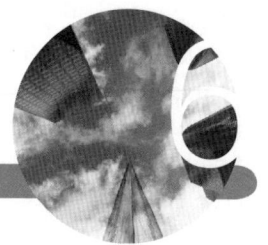

GM foods and life patents

This chapter continues the investigation of current and future food supply issues by reference to the promises and problems of genetically modified foodstuffs. It also briefly touches upon issues of life patents as 'driving forces' of such genetic modification.

The promise

In their book *Reshaping Life: Key Issues in Genetic Engineering*, distinguished biological researchers G. J. V. Nossal and Ross Coppel (2002, p. 148, pp. 152–3) argue that feeding 'increasing numbers of people on the planet' is 'probably the greatest problem facing humanity'. They refer to the 'impressive achievements' of science in increasing crop yield in past decades, including 'the green revolution which has allowed global food production to keep up with an expanding population' which 'rests on a disciplined and institutionalised, ever-changing research base' and 'the doubling of yield in the US over the last 40 years'. But they say that 'most experts believe that the conventional methods of improvement in yield of food crops have gone about as far as they can'.

As they say

> If we have exhausted conventional technology, what is there with the potential to keep pace with the increasing demand for agricultural products and protect the environment and ensure the safety, security and diversity of the food supply? Nothing other than GMOs appears able to do this. (p. 153)

Proponents of genetic modification have highlighted the significant price increases of agricultural products over the last three years (see Chapter 5), attributing such increases, in part, to productivity stagnation in face of demand from an expanding world population. Given that little cultivatable land around the world remains uncultivated, genetic modification technology is then presented as the only way to achieve what the United Nations (UN) says is a necessary 50% increase in agricultural output by 2030.

Paul Roberts (2008, p. 244) notes that previous forms of genetic manipulation (through selective breeding and irradiation) have been largely random and very slow but have still managed to increase US corn yields sixfold since 1930. Most importantly, such techniques now seem to be approaching the end of their usefulness with most possibilities of cross-breeding and many possibilities of randomly induced mutations having already been tried. The use of transgenes promises to allow a 'controlled upgrading of genetic material' to break through all such restrictions. He quotes proponents as claiming that 'plants [designed] to tolerate heat and drought and salty [degraded] soils, to use nitrogen more efficiently, to produce vastly more edible weight' and resist diseases are just around the corner. He quotes Monsanto researchers as predicting per acre yields twice as high as the currently highest yields (from 100 bushels per acre to 200 bushels) (P. Roberts, 2008, p. 245).

Other possibilities include the use of transgenic plants to produce pharmaceutical products, such as drugs and vaccines, and 'fortified foods' like golden rice, modified to produce vitamins important for human nutrition, and high energy crops for use in bio-fuels.

Nossal and Coppel (2002, p. 157, pp. 158–9) point out that 'just as with plants, it is possible to cut and paste genes and introduce them into pigs, sheep, goats and cattle'. While 'the time taken to breed' transgenic animals is much greater than with plants, so that 'commercial exploitation of GMOs [genetically modified organisms] will initially focus on plants', nonetheless there are also 'a number of interesting developments to do with animals'. These include the introduction of growth hormone genes to increase meat yield and the production of growth hormone for use in farm animals by genetically modified (GM) technologies.

> This [latter] approach is commonly applied in the US in cattle using bovine growth hormone (also known as bovine somatotropin). This product, produced

by Monsanto and called Posilac, has been available since about 1994. It is currently used in about one third of all dairy cattle in the USA, and is the largest selling dairy animal health product in the US. Its use results in increased milk production . . . leading to increased profitability for the producer . . . (Nossal and Coppel, 2002, pp 158–9)

Future possibilities include the use of animals to produce human pharmaceuticals in their milk, their use as organ donors for people (removing genes that promote rejection or inserting human genes), increased wool production, and bigger, faster growing fish. Transgenic fish, in fact, growing 'faster and bigger, more disease resistant' and 'able to digest new types of food' 'may be of immense use in helping to feed mankind' (Nossal and Coppel, 2002, p. 161).

On the subject of safety, the website of Food Standards Australia New Zealand (FSANZ), the body responsible for food safety issues in Australia (and New Zealand), assures us that

> . . . the Office of the Gene Technology Regulator (OGTR) oversees . . . development and . . . release of GM organisms under the Gene Technology Act 2000 . . . licenses will not be issued unless the OGTR is satisfied that . . . risks . . . can be managed . . . to protect . . . people and . . . the environment . . . each new genetic modification is assessed individually for its potential impact on the safety of the food. We compare the GM food with a similar . . . conventional food from a molecular, toxicological, nutritional and compositional point of view. The aim is to find out if there are any differences between the GM food and its conventional counterpart, which we already know to be safe . . . If the . . . modification causes an unexpected effect . . . such as increasing . . . allergenicity or toxicity, it will not be approved. To date, we have identified no safety concerns with any of the GM foods . . . we have assessed. Other national regulators . . . have reached the same conclusions. (http://www.foodstandards.gov.au/foodmatters/gmfoods/index.cfm)

These sorts of ideas – of the need for and the benefits and safety of GM foods – have been reiterated by GM proponents and lobby groups and government departments (suborned by the big GM transnational corporations (TNCs)) since the 1980s. The arguments of Chapter 5 have already cast serious doubt upon a number of them. In particular, doubt has been cast upon the triumphal 'wonders of science save the world' interpretation of the Green Revolution, upon the 'overpopulation creates hunger' argument and the 'no further significant improvement possible with conventional methods'.

We have seen that the Green Revolution was far from being the only possibility for addressing hunger in the developing world, that it was designed to protect inequality and privilege and avoid land reform rather than feed the poor, and that it has been largely responsible for a developing food crisis.

We have seen that it is radical inequality and poverty, created by the market system, which are responsible for hunger today, rather than overpopulation. That agricultural output can be increased and made sustainable in the future through 'conventional' technologies of labour intensive organic farming as demonstrated, for example, in Cuba. In this chapter, we consider serious reasons to doubt the claims for the qualitatively higher levels of control achieved by the new genetic technologies, for the benefits and safety of currently available GM crops, and for the imminent arrival of significantly improved crops that are safe, cheap, higher yielding, drought proof, salt proof, and nitrogen-fixing.

At this stage, we highlight the tendency of GM proponents to refer repeatedly to the 'great promises' GM holds for the future rather than to the real value of past achievements. As will be shown, this is because of the limited nature of such past achievements. We will cast doubt upon the claims for significant improvements in the future. Here, indeed, even Nossal and Coppal (2002, p. 156) acknowledge that while recent developments have indeed been 'clever', 'it is not yet possible to know which of these manipulations will eventually lead to improved crops'.

Genes

There is a widespread belief, carefully fostered by the gene engineering industry, that genes exist as discrete sequences of chemical building blocks called nucleotides. Each nucleotide is composed of a sugar molecule, a phosphate and one of the nitrogen-rich bases: adenine, guanine, thymine or cytosine. 'Thousands of nucleotides come together in pairs to form a single molecule of DNA' (Robinson, 2005, p. 83). Chromosomes are strands of DNA in the protected central 'nucleus' of the living cell, with genes as local linear sections of such chromosomes, around 3000 base pairs long, which 'make up the building plans for physical traits' (Robinson, 2005, p. 25). Through a process of 'transcription', the DNA 'plan' of the gene is copied over to a separate 'messenger' molecule of RNA, which takes it out of the nucleus into the body of the cell, called the 'cytoplasm'. 'Each gene in the DNA molecule [is] transcribed as an intermediary molecule of RNA, which is 'in turn, translated into an amino acid sequence that makes up a protein' (Rouvroy, 2008, p. 14). Such proteins are then responsible for controlling all other living processes.

According to this model, there is a one way flow of information from a specific gene, exposed through unravelling of the DNA by an enzyme (RNA polymerase), to messenger RNA (created by the polymerase as a copy of the gene), to ribosome (a protein assembly point outside the nucleus), to specific transfer RNAs (which bring amino acids to the ribosome), to a specific protein (formed through linking together amino acids into a peptide chain), which then becomes a part of the cell (Walker, 2007, pp. 44-7). The theory is that DNA 'contains the complete information that defines the structure and function of the organism', with each single gene producing a particular single protein (Rouvroy, 2008, p. 14).

This model has had profound consequences in many areas. It has encouraged and justified substantial public and private funding of research, including the rapid mapping of the human genome, which could have been used in other areas of health and welfare provision, on the grounds of such research being the basis for substantial and rapid progress in medical science. It has encouraged and justified the idea of gene patenting, on the basis of the social benefits of rewarding the costly, but ultimately straightforward, work required to identify the particular individual functions of individual genes as a basis for genetic testing and genetic engineering.

Even those who have acknowledged an inevitable interaction of environmental factors with the genome to produce particular consequences in terms of adverse health outcomes or desirable new phenotypical traits, have argued that because of the difficulty and complexity of such environmental influences, 'the main causal factors should generally be seen as genetic' (Rouvroy, 2008, p. 48).

As has frequently been pointed out, such genetic reductionism and determinism – identifying genetic predispositions as the main 'causes' of adverse health outcomes – tends to absolve those in a position to influence environmental determinants of such outcomes of any legal or ethical responsibility for them. Genetic predispositions to particular cancers, for example, would not have produced such cancers in the absence of particular environmental carcinogens. Such toxins could relatively easily have been removed from working and living environments by employers or government agencies. But people with the relevant gene are thought of as intrinsically disease prone, rather than being potential victims of environmental vandalism.

On the other side of the coin, genetic engineers are seen as wholly responsible for any desirable changes produced by their genetic manipulations without regard to the necessary involvement of other genes, organs and particular environmental factors or the huge background of practice and research underlying such manipulations.

The Human Genome Project found only 25–30 000 entities identifiable as genes, only around one third more than a roundworm and about the same number as mustard weed, even though humans are estimated to make as many as 300 000 different types of protein (Walker, 2007, p. 50). This showed that the role of genes is rather more complex than the 'orthodox' reductionist and determinist picture suggests (Walker, 2007, p. 22), that we need to consider the 'complex dynamic molecular networks involving relations between genes, enzymes, non-cellular structures and environmental factors' in order to develop effective technology of genetic diagnosis, treatment and transformation rather than merely correlating individual genes with individual proteins and individual illnesses.

It turns out that the effects of particular genes depend on which other genes are present, that genes have multiple functions and 'convey many different and often unrelated messages' with 'the timing and nature of the messages not exclusively determined by the genes themselves' but by the non-genetic condition of cells, organs and the body itself.

> Each gene can be used in a variety of different ways, depending on how it is regulated... Furthermore, genes are made up of sub-units that can be put together and combined with parts of other genes in a variety of different ways. A mutation in one gene will have several potential phenotypical manifestations. Conversely more than one gene may command a given function.
> (Walker, 2007, p. 24)

The same disease can be a consequence of different genetic structures, and genetic tests generally do not provide definitive evidence for the development of any particular disease condition.

The fact that particular pieces of DNA 'need to interact in a complex manner with DNA in other parts of the genome, enzymes, non-cellular structures and the environment in order to produce one or several proteins' (Walker, 2007, p. 26) means that the traditional concept of the gene has been seriously undermined. As Rouvroy points out, the fact that individual genes actually have multiple functions and convey many different messages means that when some researchers are granted a patent on a particular gene, supposedly associated with a particular function, they can then restrict and 'tax' further research into the wider functional significance of the gene in question and its components in interaction with other genetic and environmental conditions (Rouvroy, 2008, p. 42).

As far as genetic engineering of food crops is concerned, these considerations highlight the dangers of rapid commercialism of particular modifications without adequate knowledge of the inevitably complex and multiple consequences of such changes by virtue of the interaction of gene components, cells, organisms and environments.

GM crops

Monsanto, DuPont, Syngenta, Bayer CropScience and Dow are the main controllers of GM seed technology. In 2005, Monsanto technology accounted for nearly 90% of GM acreage, with these five companies having 35% of the world seed market and 59% of the world pesticide market. As Jeffrey Smith notes, the four principal GM crops are soybeans, corn, canola and cotton.

> All are used to make vegetable oils and soy and corn derivatives are used in most processed foods. There are also GM zucchini, squash, papaya, alfalfa and tobacco. GM tomatoes and potatoes were taken off the market. (2007, p. 7)

The main GM trait is herbicide tolerance, with herbicide tolerant (HT) crops engineered to survive exposure to high levels of weed killer. Farmers have to buy the companies' own brands of herbicide along with their HT seeds, including Monsanto's Roundup (glyphosate) and Bayer's Liberty (glufosinate ammonium). These are 'broad spectrum herbicides designed to kill all other plant life' (Smith, 2007, p. 7). Roundup ready crop varieties contain a gene derived from bacteria found to survive in chemical waste ponds near the Roundup factory. They allowed Monsanto to maintain its herbicide sales after its patent ran out in 2000, with farmers being forced to buy Monsanto's brand of glyphosate along with the seeds, and typically using increasing amounts of it with the HT crops (Smith, 2007, p. 7).

Monsanto has sued many US farmers for saving and replanting seeds that contain the company's patented transgenes without paying a 'technology fee' and they have worked on terminator seeds for the future.

The second major trait is insect resistance. A gene from the soil bacterium *Bacillus thuringiensis* (Bt), is inserted into crop DNA, producing insect killing toxins in every cell of the plant. Sixty eight per cent of GM crops are designed to resist a herbicide, 19% produce their own pesticides, and a few contain modified viral genes which resist infection by particular plant viruses (e.g. Hawaiian papayas) (Smith, 2007, p. 7).

Six countries grow most commercial GM crops: the US (57.7 million hectares in 2007), Argentina (19.1 million hectares), Brazil (15 million hectares), Canada (7 million hectares), India (6.2 million hectares) and China (3.8 million hectares) (followed by Paraguay, South Africa, Uruguay and the Philippines). Such GM crops are also grown in other countries including Mexico, Spain, the UK and Australia (Smith, 2007, p. 7).

In 1998 a temporary moratorium was imposed by the European Union (EU) upon imports of any foods or seeds containing transgenes, to allow for further scientific study of such products. In 2003 the US, along with Canada and Argentina, filed a World Trade Organization (WTO) complaint against the EU's GMO moratorium, maintaining that it was a 'barrier to trade' costing international producers hundreds of millions of dollars of lost sales (Schapiro, 2007, p. 94). Following discussion in the Dispute Resolution Panel in Geneva in 2005, the WTO declared the ban 'an illegal barrier to trade'. By then the EU policy had changed, replacing the blanket ban with a case by case assessment process, putting particular genetically modified organisms through scientific review.

As in Australia, such review requires the submission of environmental and health data to a regulatory body, the European Food Safety Agency, for consideration prior to approval for commercial cultivation. By early 2007, 17 varieties had already been approved for cultivation in Europe, subject to the use of buffer zones distancing genetically engineered (GE) crops from neighbouring non-GE crops to try to prevent cross-pollination. All GMO containing products have to be labelled, which will mean rejection by the majority of European consumers. Retail outlets across Europe have committed themselves to selling no GE goods, and approved crops will probably mostly be used as cattle feed. One hundred and sixty six regional governments in Italy, Poland, Austria and other countries 'have instituted safety provisions so strict that they are, in effect, GMO free'. Some regions, including Brittany, have organised special trade deals to keep GMO products out.

> In December 2006 19 of the EU's 25 member [states] voted against the Commission's request that Austria lift its ban on 2 GE corn varieties...
> (Schapiro, 2007, p. 97)

In the US, organic farmers' livelihoods have already been destroyed by pollution of their crops by neighbouring GE crops. In France, farmers' action groups have destroyed GE crops and lower courts have supported these actions as a defensive response to 'a situation of necessity' (Schapiro, 2007, p. 97). An EU *Environmental Liability Directive* now requires member states to 'include provision in their national laws offering compensation to farmers whose crops are contaminated by gene flow from GMOs' (Schapiro, 2007, p. 97).

All canola growing states in Australia imposed bans on commercial herbicide tolerant genetically modified (GM) canola in 2003. The Victorian and New South Wales (NSW) Governments have since allowed their bans to lapse (on 29 February and 3 March 2008, respectively). As Bob Phelps (2007 – personal correspondence) notes, Premier Brumby in Victoria allowed the planting of herbicide tolerant GM canola from Bayer and Monsanto

> ... without restriction or public notice... NSW... extended its ban till... 2011 but will exempt some GM canola growing for commercial and research purposes... An expert panel will advise the Agriculture Minister... The majority of citizens... want the bans extended... Several state governments also want to remain GM-free but GM canola pollen and seed will cross state borders.

A number of food processors and retailers in Australia, including Heinz, Pureharvest, McCain, Weet-Bix, Campbell's and Bega, have committed themselves to trying to keep transgenes and their proteins out of their products. Others, including Woolworths, Kraft, Nestlé, Birds Eye, Bakers Delight, Maggi and MasterFoods, have made no such commitment (Greenpeace Australia, 2009).

Making GM crops

Codes in genes from bacteria have to be modified to function in plants. Promoter sequences are added at one end to switch the gene on and force the gene to continuously produce a particular protein. The all-purpose promoter most often used for transgenes in plants comes from the pathogen called cauliflower mosaic virus (CaMV) (Smith, 2007, p. 9).

A terminator sequence is added to the other end to tell the plant DNA where the transgene stops, along with an antibiotic resistant marker gene. This so-called 'gene cassette' is put into a circular piece of bacterial DNA called a plasmid, which can copy itself within a bacterium to make numerous copies of the cassette (Smith, 2007, p. 9).

The plant cells, which will receive the transgenes, are grown in tissue cultures that tend to produce numerous uncontrolled mutations in their DNA. Two main methods are then used to put the new genes into the plant cells. The first uses a common soil bacterium called *Agrobacterium*, which infects plants through wounds by inserting a portion of its own DNA into the plant DNA. This causes the plant to grow tumours called galls. Genetic engineers replace the tumour creating section of the bacterial DNA with one or more other genes. The modified bacterium then infects the plant cells' DNA with these foreign transgenes. As the plant cells are totipotent (capable of generating all plant cell types), they can grow into entire plants with the transgene in every cell, and in every seed the plant produces (Smith, 2007, p. 9).

Some plants are resistant to *Agrobacterium*, and require the use of a second, less reliable method. In this case, the genetic engineers coat millions of particles of tungsten or gold with gene cassettes and then shoot these, using compressed air, into millions of plant cells. Only a few of these cells incorporate the foreign gene (Smith, 2007, p. 9).

The engineers rely on marker genes to select cells that have successfully integrated the gene cassette. Antibiotic-resistant markers are designed to confer resistance to a particular antibiotic that would otherwise kill the cell. That antibiotic is applied to the cells after the gene insertion process. Those that survive are the ones that have the marker gene operating in their DNA. Most cells die. The surviving cells are again grown using tissue culture but with a changed nutrient medium which allows them to develop into plants.

The engineers can plant the seeds from these plants or make clones through tissue culture of the plants cells to multiply the desired transgene.

Each plant grown from a separate gene insertion is unique with unique genetic properties, never previously existing, and the strains derived from each insertion are called 'events'. The whole process is highly mutagenic, and leads to the production of foreign proteins at high levels in all parts of the plants which are produced.

Proponents of GE have presented the process as a controlled introduction of known elements into known locations with known – and limited – results. In fact, the process is largely uncontrolled and can produce massive unforseen changes in the natural functioning of the plants' DNA.

> Native genes can be mutated, deleted, permanently turned off or on, and hundreds may change their levels of expression. The inserted gene can become truncated, mixed with other genes, inverted or multiplied and the GM protein it produces may have unintended characteristics with harmful side effects. (Smith, 2007, p. 3)

Transgenes can produce quite different proteins from those intended. Intended proteins can behave in new and harmful ways. For example, with the development of GM peas in Australia, a protein supposedly identical to a harmless natural one caused inflammatory responses in mice, and could have triggered deadly allergic responses in people (Smith, 2007, p. 4).

Even where the protein structure is precisely as intended it can have harmful consequences. Bt toxin produced by GE corn and cotton can produce a range of allergic type symptoms and is implicated in a growing number of human and livestock illnesses. Rats fed Monsanto's MON 863 Bt corn for 90 days 'showed significant changes in their blood cells, livers and kidneys, which might indicate disease' (Smith, 2007, pp. 26–7). Mice fed GM Bt potatoes developed possibly precancerous intestinal damage (Smith, 2007, pp. 28–9).

Inserted genes may transfer from food into gut bacteria or internal organs. As Smith notes

> the only human feeding trial ever published confirmed that genetic material from Roundup Ready soy transferred into the gut bacteria in three of seven human volunteers. The transferred portion of the transgene was stable inside the bacteria and appeared to produce herbicide tolerant protein. (2007, pp. 130–1)

Animal studies show that inserted DNA can travel throughout the body from GE foods, even into a foetus via the placenta. Critics highlight the possibility of Bt toxin genes becoming established in gut bacteria, turning such populations into 'living pesticide factories', threatening human populations in high corn consumption areas like Mexico and South Africa, with chronic poisoning and life threatening allergy (Smith, 2007, p. 4, p. 137).

Regulators and biotechnology companies are simply asserting that this won't happen without any proper studies to support such a claim.

> The gene that codes for Bt has proven stable in the presence of saliva and within the digestive tract of animals. The CaMV promoter, which drives the expression of Bt genes is active in bacteria. There may be little standing in the way of Bt gene transfer. (Smith, 2007, p. 137)

Doctors in many countries have highlighted the danger that antibiotic resistance marker genes may transfer from GM foods into bacteria in the human digestive system or in the environment, 'spread to pathogenic bacteria and create antibiotic resistant diseases'. The Bt 176 corn variety by Syngenta carries an ampicillin resistance marker, even though ampicillin is a widely used antibiotic and the drug of choice for several types of human and animal infections. Most GM crops use kanamycin resistance markers, with regulators arguing that it isn't much used anymore. However, it remains a valuable antibiotic in a number of situations and kanamycin resistant bacteria can mutate into forms resistant to a range of other related antibiotics (Smith, 2007, p. 133).

Contrary to prior assumptions it has been found that the CaMV promoter, used in nearly all GM crops, does function in human, animal and bacterial DNA. This promoter may transfer into the DNA of human gut bacteria and might also transfer into human DNA. Once transferred it may switch on genes that produce toxins, allergens, or carcinogens, create genetic instability and, in higher organisms, switch on dormant viruses.

Bayer's Liberty crops are engineered to withstand glufosinate based herbicide. The crops transform the herbicide into a compound regarded as non-toxic, called NAG, which remains in the plant. Once humans or animals consume NAG, gut bacteria can revert some NAG back into toxic herbicide, which can kill off or disturb gut micro-flora. If the herbicide tolerant gene transfers to gut bacteria, it could magnify these problems (Smith, 2007, p. 145).

Herbicide tolerant crops increase the use of their associated herbicides. Increased herbicide residues in crops can promote the toxic effects of these chemicals on humans, animals and their offspring. They can alter nutrient content, such as flavonoids, making GM crops less nutritious.

There is now massive evidence of the escape of transgenes into other hosts. Just two years after the first sales of herbicide resistant canola in Canada in 1996, two different transgenes for herbicide resistance were found in wild canola plants in fields without any transgenic crops. By 1999 the first triple resistant canola was discovered. Walker (2007, p. 225) notes, 'canola is an open pollinating plant' and development of such multi-resistance 'in a related weed population, wild mustard, would be disastrous'.

As noted above, organic farmers in the US have already been ruined by genetic pollution of their crops.

> In Spain, the only country in Europe with large scale commercial growth of GMO crops [limited to use as cattle food], Greenpeace reported that '25% of non-GE corn samples from Aragon and Catalonia had traces of GMOs'. [In 2006 the GMO Contamination Register] listed dozens of contamination events... in 25 countries involving 'corn, rice, soybeans, papaya, rapeseed and canola'. (Schapiro, 2007, pp. 88–9)

While Quist and Chapela's original 2001 report of transgenes from GM corn widely distributed in traditional, wild corn varieties in Oaxaca, Mexico, faced a barrage of criticism from many different directions, *New Scientist* (http://www.newscientist.com/article/mg20126964.200-alien-genes-escape-into-wild-corn.html) recently reported solid confirmation from investigations by Elena Alvarez-Buylla of the National Autonomous University in Mexico City and her team who found transgenes in about 1% of nearly 2000 samples they took from the region (Piñeyro-Nelson et al., 2009, p. 750). The jury is apparently still out on Quist and Chapela's original claim that the introduced DNA has integrated into the wild grains at multiple random sites, thereby 'turning on genes not normally active and deactivating necessary genes' (Walker, 2007, p. 226).

The accelerated emergence of herbicide resistant weeds has resulted in the increased use of yet more toxic varieties of herbicides. As Paul Roberts notes

> once super-weeds emerge – and since RoundupReady crops were introduced in 1996, thirteen weed species in fourteen [US] states have become resistant to glyphosate – farmers must find a new herbicide, and in some cases they've been switching to older, more potent [and much more toxic] products, such as paraquat and 2,4-D. (P. Roberts, 2008, p. 256)

The 'escape' of genes engineered to reduce the rigidity of wood (by interfering with lignin biosynthesis) in plantation trees (so as to cut the costs of paper production) into wild tree populations would have disastrous consequences (Walker, 2007, p. 226). Lawn grasses designed to be herbicide resistant and needing less water 'could invade diverse types of habitat' while being impossible to kill (Walker, 2007, p. 192).

Gene flow is of particular significance in relation to possible future transgenic plants designed to manufacture pharmaceutical compounds. Potatoes, grains, bananas and rice are already being used to produce monoclonal antibodies, vaccines and digestive enzymes. As Paul Roberts (2008, p. 257) notes, this raises the possibility of such crops mating with crops grown for human consumption 'with the unanticipated result of novel chemicals in the human food supply'. Indeed, he quotes an executive of Pfizer Pharmaceuticals admitting that such crossing has already happened in experiments involving vaccine producing plants. In 2003 the US Department of Agriculture (USDA) animal and plant inspection service introduced new, more stringent rules for field testing of such plants. However, contamination of food and feed crops seems inevitable.

An issue touched upon in Chapter 5 was that of an increasing shift to monoculture associated with GM crops. As noted there, GM crops continue a well-established trend towards fewer crop varieties, rendering crops vulnerable to disease and environmental change. Insecticide producing GM crops discourage traditional seasonal crop rotation to control particular insect pests.

Insect resistant crops kill beneficial insects. Resistance develops amongst insect populations when insects that are susceptible to the pesticide transgene are all killed. The only insects which survive and reproduce are able to tolerate the pesticide transgene. Organic farmers have been particularly concerned that developing Bt resistance deprives them of a relatively safe organic pesticide. They could continue to use Bt toxin themselves because of the survival of non-resistant insect strains to replace those killed by the toxin. With vast acreages of GE crops producing large amounts of the toxin in all their tissues the danger of resistance is hugely increased.

The US Environmental Protection Agency (EPA) now requires farmers using Bt engineered crops to reserve 20% of their acreage for 'traditional crops' as 'refuges' for Bt susceptible insects. In 2000 30% of US farmers were found to be ignoring the protocols' (Walker, 2007, p. 226) and resistance could anyway be a dominant trait leading to breeding out of susceptibility.

The insertion of viral transgenes into GM plants carries the risk of their recombining with natural viruses in the same plant to create new offspring viruses, 'these transgenes produce viral proteins which may be toxic or suppress viral defences in humans'. If they enter the DNA of gut bacteria they could 'produce viral proteins (and RNA) in the gut over the long term' (Smith, 2007, p. 141).

The few studies of toxicity of GM foods that have been carried out show laboratory animals fed GM foods suffer stunted growth; impaired immune systems; bleeding stomachs; abnormal and potentially precancerous cell growth in the intestines; impaired blood cell development; misshapen cell structures in the liver, pancreas, and testicles; altered gene expression and cell metabolism; liver and kidney lesions; partially atrophied livers; inflamed kidneys; less developed brains and testicles; enlarged livers, pancreases and intestines; reduced digestive enzymes; higher blood sugar; inflamed lung tissue; increased death rates and higher offspring mortality (Smith, 2007, pp. 168–72).

An increasing number of farmers report the GM corn varieties caused their pigs or cows to become sterile, that their sheep died from grazing on Bt cotton plants, and that cows, buffaloes, chickens and horses have died from eating

GM crops. Biotechnology proponents argue that millions of people around the world, particularly in the US, have been eating GM foods for more than a decade with no ill effects. Critics point out that the HIV epidemic went unnoticed for decades, and that without much more investigation, we don't even know where to look for possible new health problems produced by GM foods. By the time they do show up, potentially a huge amount of damage will already have been caused.

Even with 5-10,000 sick and about 100 dead, the epidemic caused by the GM food supplement L-tryptophan was almost missed. It was only discovered because it was unique and acute. The health consequences of other GM foods could easily be more chronic and long-term and could be missed for a long time.

Epidemiologist Judy Carman refers to the experience with the tobacco industry, where millions had already died before definitive evidence of toxicity became available. Even then the industry responded with lies, threats, obfuscation and misuse of the legal process while millions more died. In this case politicians could also be influenced by thousands of farmers whose living could be threatened. 'So, even if a GM food is found to cause harm, it may take many years of effort to remove it from the food supply' (Smith, 2007, p. 11).

Containment problems are highlighted by the case of the Bt corn variety called StarLink, approved for sale only as an animal feed, with StarLink Bt toxins found in 300 grocery items, including Kraft taco shells, probably as a consequence of major difficulties in segregating different product streams in modern production and distribution processes (P. Roberts, 2008, p. 255).

As Roberts also notes, this shows that 'even those producers or consumers who wanted to shun the gene revolution might not have a choice in the matter'. And

> even today, consumers [in the US] wishing to avoid transgenic foods cannot, because the industry has successfully blocked any requirement that transgenic foods be labelled – despite surveys showing that 9 out of 10 consumers want such labels. (2008, p. 256)

It has been estimated that already, by 2001, 60% or more of processed foods on US supermarket shelves contained at least one GM component, usually soy, canola or corn. US companies have spent substantial sums of money successfully opposing campaigns in a number of US states for labelling of GM foods.

In Australia, it has been mandatory for GM foods to be identified on food labels since 2001. Due to exemptions allowed under the labelling requirements, virtually no foods are actually labelled in practice. The regulations exclude highly processed food where the processing removes all DNA and/or protein; and minor ingredients, including processing aids and food additives (unless they contain novel DNA and/or novel protein. Such labelling requirements also allow a food in which an approved GM food is unintentionally present in a quantity of no more than 10g/kg (1%) per ingredient to remain unlabelled.

FDA issues

As Paul Roberts points out, in the US, oversight of transgenic foods is split amongst three agencies

> The USDA oversees only the crop trials for proposed transgenic foods. The EPA regulates the pesticides produced by transgenic crops (such as Bt in Bt corn) but looks only at the gene and the gene products, not any potential health impacts; these... are left to the US FDA. But unlike European regulators who treat [trangenes] as food additives and subject them to mandatory testing before products reach the market, the FDA does not test transgenic foods prior to their release. (2008, p. 255)

The US Food and Drug Administration (FDA) has played a key role in spreading the idea that there is solid evidence for the safety of GM foods. In 1992 the FDA stated that it was 'not aware of any information showing that foods derived by these new methods differed from other foods in any meaningful or uniform way'. On the basis of this pronouncement the FDA claimed that no safety studies are necessary and that 'ultimately it is the food producer who is responsible for assuring safety' (Smith, 2007, p. 1). Biotechnology companies thus determine themselves that their products are harmless.

Defenders of agricultural biotechnology picked up on this authoritative statement. They pointed out that in addition to thousands of years of selective breeding, people have, for over 70 years, exposed plants to radiation and mutagenic chemicals to produce mutant alleles hopefully associated with desired traits. Fruits, vegetables and grains are mutated to produce disease resistance, flavour variations and changed times of fruiting. At the same time viruses and bacteria are engaged in continuously moving genes from one species to another, as do human genetic engineers.

The problem is that none of these things are really comparable with contemporary genetic engineering. Now, transgenic organisms end up with genes that could not have been produced through selective breeding or intentional – random – mutation, and never could have moved from one organism to another without a lot of help. Government scientists in the FDA were well aware of this.

Jeffrey Smith observes that FDA policy, based upon the idea of 'equivalence', allowed for rapid development of the new technology, with large-scale planting, contamination and consumption. But in fact

The . . . consensus among the . . . experts in the agency was that GM crops were meaningfully different . . . [with] different risks than traditional breeding . . . [and] unpredictable, hard-to-detect side effects. They urged the political appointees . . . in charge of the FDA to require long-term safety studies, to guard against possible allergies, toxins, new diseases and nutritional problems. The scientists' concerns were kept secret in 1992 when FDA policy was put in place . . . 7 years later internal records were made public due to a lawsuit and the deception came to light. (2007, p. 1)

The records showed deletion of government scientists' warnings. They showed that the FDA was under orders from the White House to promote GM crops and that Michael Taylor, Monsanto's former attorney and later vice president, was brought into the FDA to oversee policy development. As a result consultation with the FDA on GM food safety is a voluntary exercise and if the company claims that its foods are safe the FDA has no further questions.

In the mid 1990s the UK Government instituted rigorous long-term safety testing of some GE crops. Three years into the project the scientists discovered that potatoes engineered to produce a harmless insecticide caused extensive health damage to rats. The pro-GM government immediately cancelled the project, the lead scientist was fired and the research team dismantled.

Policy makers around the world gain confidence in the safety of GM crops because they wrongly assume that the US FDA has approved them on the basis of extensive tests. The USDA spent US$250 million helping develop and promote agricultural biotechnology. However, less than 1% of this [$1.6 million] was spent on risk assessment (Schapiro, 2007, p. 90).

FSANZ

In Australia, it is the FSANZ that is responsible for reviewing the safety of GM food. As noted earlier, this organisation 'carries out safety assessments on a case by case basis' and it issues reports of about 70 pages per application. The FSANZ does not carry out its own scientific studies, but rather relies on the 'companies that have developed GM foods to demonstrate the safety of that food and to supply FSANZ with the raw data'. The organisation does 'not routinely' require any animal feeding studies.

> FSANZ considers that a scientifically informed comparative assessment of GM foods with their conventional counterparts can generally identify any potential adverse health effects [and] for most GM foods, animal studies are unlikely to contribute any further useful information. (Smith, 2007, p. 2)

This raises interesting questions of where all the other 'useful information' is actually coming from. Smith reports that it is routine for FSANZ reports to give no actual data on experiments, and those that do provide data indicate no feeding trials on people, only a single (often very low) dose to an animal followed by 7–14 days' observation or no animal testing at all. Such observations focus 'only on the substance that has been genetically engineered to appear [the GM protein] and not on the whole food'. Internal documents from this body indicate that it 'considers that GE food is safe until proven unsafe' (Smith, 2007, p. 183).

Feeding the hungry

Hundreds of millions of dollars have been spent by biotechnology companies trying to convince people that GM crops are needed to feed the world. Their message targets those in both developed and developing countries in order to create an impression that it is morally wrong to oppose the technology.

The claims are that GM crops are safe, produce consistent yields, higher than those of non-GM alternatives, and that such higher yields are what is needed to address current food shortages in the developing world. As Smith points out, none of these claims are actually true. As we have already seen, there are good grounds for believing GM crops to be far from safe. The current crops have not been produced to achieve higher yields and generally don't do so; their yields can be dangerously inconsistent.

Higher yields will not be easy to achieve with future generations of GM crops. Roberts quotes Kendall Lamkey, a breeding expert and at that time chair of Iowa State University's agronomy department, to the effect that

> the massive yield gains the industry promises will require a level of technical mastery that . . . transgenic technology may be hard-pressed to deliver. Whereas existing transgenic successes . . . such as herbicide tolerance involve the manipulation of just one or two genes, yield . . . involves so many different traits and underlying chemical processes that so far, most of the big yield gains researchers have engineered trans-genically hold up only in the laboratory or in a narrow set of field conditions. (P. Roberts, 2008, pp. 257–8)

Even if such gains could be maintained in field conditions, past experience suggests that the crops in question will be dependent upon high levels of oil and gas based chemical assistance and water, contributing to global warming, wastage of water and soil degradation, and simply not sustainable in the longer term.

In light of the issues raised in Chapter 5 of sustainability, diversity and scale, other methods are far better for improving yields and improving the lives of developing world farmers. Increasing crop productivity will not, in itself, eradicate hunger. What is required is radical reform of land ownership, and planned and fair distribution, rather than profit driven, market chaos.

Similarly, salt resistant GM grain crops sound, at first, like a solution to the salinity crisis considered in Chapter 5. But this ignores the principal cause of the crisis, in the form of rising water tables. As Beresford et al. (2004, p. 247) point out, most of such salt tolerant cereal crops produced so far have failed to survive because of waterlogging.

They argue that the ecological devastation of serious salinity means that 'using salt affected land productively should be reserved for those areas where alternatives lack feasibility' (Beresford et al. 2004, p. 246). We need to restore a balanced and sustainable ecology not monocultures of salt resistant wheat grown in barren salt deserts or swamps.

Research is still needed to work out the best ecologically regenerative and sustainable solutions in different affected regions. Reforestation of 20-80% of catchments with original perennial trees, shrubs and understory plants to absorb rainwater before it seeps into groundwater, could be compatible with agricultural crop rotations in some areas. This issue is considered in more detail in Chapter 7.

Ethical issues

As Bremner observes, current technologies for making GM foodstuffs are essentially the same as for making GM drugs and the risk of creating a serious unexpected hazard is comparable. Food is consumed in far greater quantities for longer periods than drugs so food needs to be safer. Yet testing of genetically modified foodstuffs has been far less stringent. The justification for testing drugs on animals and volunteer humans rests on their potential for relieving pain and suffering and saving lives. Yet no one can claim that any of the GM products currently on offer or in prospect in the near future will be life savers (Bremner, 1999, p. 22).

As we have seen, the 'necessary to feed a starving world' arguments carry little weight. Since starvation is an issue of inequality and poverty, and since GM crops offer little prospect of massively increased and sustainable yields, cheaper prices or poverty reduction in the developing world, there is little reason to take any such claims seriously.

GM crops could be seriously harmful, so they cannot ethically be widely planted and sold without comprehensive safety tests on animals and humans comparable to or much better than those used to test drugs. Ethical tests cannot be carried out because there is no justification for exposing people to health risks needlessly (where the potential benefits do not significantly outweigh the risks). Since current GM food isn't needed, there is no justification for these tests. Therefore, in their absence, all GM crops should be banned immediately.

GE animals

Genes can be inserted into an animal's genome during the process of fertilisation. When a sperm enters an egg there's a brief period before the two sets of DNA – paternal and maternal – fuse to become one. The two sets of DNA existing during this intermission are called pronuclei. Geneticists discovered that by injecting many copies of the transgene with its promoter and sometimes with a marker gene directly into the paternal pronucleus, the transgene was sometimes integrated into the embryo's chromosomes. Eggs can also be injected with transgenes after the pronuclei fuse, but the uptake of the transgene is less efficient (Robinson, 2005, pp. 294–6).

As Robinson points out, only some of the embryo's cells contain the transgene and transgenes are inserted in the chromosomes at random.

> The resulting, partly transgenic [animal] is called a chimera or a mosaic. Mosaicism is the expression of genes in some but not all cells of a given individual... To get a fully transgenic animal, many chimeras are mated in the hope that homozygous transgenic offspring will be produced from one or more matings. After researchers obtain homozygotes [chromosomes with identical transgene alleles at particular loci – ensuring the expression of the trait], they isolate the transgene line so that no heterozygotes are formed by mating transgenic animals with non-transgenic animals. (2005, p. 295)

As Robinson (2005, p. 295) notes, early experiments involving the insertion of rat, human and bovine growth hormone genes into mice produced larger mice than normal and encouraged similar experiments involving the insertion of growth hormone genes into livestock in the hope of producing faster growing, leaner animals. However, things did not pan out as hoped, with transgenic pigs only growing faster if fed high protein diets, females sterile and all prone to 'muscle weakness, arthritis and ulcers'. Similar problems have been found with cows. As yet, 'no commercially viable transgenic cows or pigs engineered for growth have been produced'.

Transgenic salmon, engineered to express growth proteins all year round, were found to grow six times faster than non-transgenic versions. They convert their food to body weight more efficiently, 'meaning that less food makes a bigger fish'. Given their high levels of aggressiveness and the inevitability of their escape from pens in large bodies of water, as with currently farmed salmon, it seems that farm raised transgenic salmon could pose a serious threat 'to natural populations of both their own and other fish species as well' Robinson (2005, p. 295). This has not prevented moves to engineer other fish for such year round growth, including trout, tilapia and turbot.

It certainly seems to be the case that, in light of the major problems of the current livestock model in meeting present and future demand for protein, serious consideration needs to be given to massive extension of fish farming, insofar as 'fish are inherently efficient feed converters'. As Roberts notes, there are major problems with current practices – including issues of 'sewage, heavy reliance on antibiotics and the unsustainability of feed supplies for carnivorous fish such as salmon'. Rather than turning to transgenic monster salmon, we need, instead, to move towards 'deep water or open water aquaculture and new plant based feed supplies' for multiple different, existing ocean species (Roberts, 2008, p. 311).

Nossal and Coppel's comments on Posilac were quoted at the beginning of the chapter. Monsanto created recombinant bovine growth hormone by inserting cow genes into bacteria. It was approved for use in the US in 1993 and used by 22% of the nation's dairy cows by 2002. 'It was also used in South Africa and Brazil but banned in the EU, Canada, Australia, New Zealand and Japan' (Smith, 2007, p. 157).

There are serious questions in relation to the approval process in the US.

> One FDA scientist said he was fired after expressing concerns about insufficient data in the rbGH analysis. Other FDA employees claimed... that a Monsanto researcher turned FDA employee raised allowable levels of antibiotics in milk one hundredfold to pave the way for rbGH approval. And when Canadian government scientists analysed how the FDA approved rbGH they found '... no critical analysis of the quality of the data' provided by the manufacturer. 'Such possibilities... as sterility, infertility, birth defects, cancer and immunological derangements were not addressed'. The Canadian scientists also testified before the Senate that Monsanto offered them a bribe of $1–$2 million to approve the drug. (Smith, 2007, p. 157)

As Smith explains

> dairy products from treated cows carry several health risks, the most serious of which are higher levels of the hormone insulin-like growth factor 1 (IGF-1). (Smith, 2007, p. 157)

This chemical is present in the milk of untreated cows and drinking cows' milk elevates human levels of the hormone by a significant percentage. Milk from cows treated with recombinant bovine growth hormone (rBGH) can have 10 times higher levels of insulin-like growth factor (IGF-1) than that of untreated cows, and has a correspondingly greater potential to raise human levels. It is not completely destroyed by the digestive process, as Monsanto originally claimed.

IGF-1 causes cells to divide, and elevated levels have been associated with greater likelihood of prostate, lung, colon and breast cancer in humans. Side effects for the treated cows include cystic ovaries, uterine disorders, decreased gestation periods, decreased calf birthweight, and increased twinning (Smith, 2007, p. 157).

> Udder infection – mastitis – is the most widely reported. This painful disease increases the pus in milk; milk from treated cows has 19% more [pus]. To manage infections and pus levels, farmers using rbGH typically treat their herd with extra antibiotics, which increases antibiotic residues in milk. (Smith, 2007, p. 157)

Milk from treated cows has higher levels of bovine growth hormone and thyroid hormone than that of untreated cows and is of lower nutritional quality. Use of rBGH has also encouraged the use of high protein feed stuffs, including rendered cow and sheep carcasses, increasing the dangers of prion infection. Yet Monsanto tried to sue farmers seemingly disparaging Posilac injections to increase milk yields of dairy cows.

There have been few 'successes' in other areas. Some researchers hold out hope for the Enviropig, developed in Canada. It produces phytase in its saliva which reduces phosphate excretion in the animal's waste. Walker argues that 'widespread use of these animals could significantly reduce the environmental impact of intensive pig farms' which, as we saw, are major polluters, particularly in the US (Walker, 2007, p. 197). However, this assumes continued criminal mistreatment of such 'factory' pigs.

In Australia, the Commonwealth Scientific and Industrial Research Organisation (CSIRO) devoted resources to attempting to produce cows with GM gut bacteria enabling them to thrive on the poor food of Australia's semi-arid regions. Yet the thin soils of these regions are destroyed by cattle and the CSIRO was, indeed, simultaneously campaigning to protect Australia's vulnerable soil and biodiversity.

Another such problematic case is that of the genetic modification of a bacterium which causes ice crystals to form on crops. Fears were raised that a genetically modified version, engineered to prevent such ice formation, could escape into clouds and change the climate through preventing rainfall (Walker, 2007, p. 196).

Life patents

Underlying the development and marketing of GMOs are legally established genetic property rights. The technologies would not have been developed in such destructive ways without the granting of patents for genes, GMOs and genetic manipulation. Patents are issued by government departments: in the US it is the Patent and Trademark Office; in Australia it is IP Australia 'helping Australian Business to prosper' as they say on the website. A patent gives the patent owner exclusive rights to manufacture and sell their invention for a certain length of time – now usually 20 years.

In the past, patents have been justified as incentives for human inventions. As Krimsky points out, Article 1, Section 8 of the *US Constitution* gives

Congress powers to grant a 'limited time' of exclusive rights to discoverers to 'promote the progress of science and useful arts'. Thomas Jefferson argued that inventors should be given entitlement to the profits arising from their inventions for a limited period in exchange for making those discoveries, thereafter, freely available to everyone, as an incentive for further useful innovation in the future.

> According to the US Patent Act [35 USC 101] a patent may be awarded to 'whoever invents or discovers any new and useful process, machine, manufacture, or composition of matter, or any new and useful improvement thereof'. (Krimsky, 2004, p. 59)

This appears to preclude 'products of nature', which clearly have not been invented, created or improved by human intervention. But as Krimsky points out, in the interests of 'greasing the wheels of progress', the US Patent and Trademark Office has, in fact, interpreted the rules sufficiently loosely to allow patents for living organisms and natural substances, insofar as such organisms were part of newly developed processes of some kind well before the issue of gene patents arose. Before the first gene patents, a number of patents were issued for micro-organisms, insofar as such organisms were involved in newly developed processes of production of useful materials.

> During the 1920s plant breeders lobbied Congress to gain the benefits of patenting that had been granted to innovations in the mechanical arts. In response, Congress passed the Plant Patent Act of 1930, which extended the definition of patentable material to certain varieties of asexually produced plants... propagated by... grafting and budding... (Krimsky, 2004, p. 63)

As Krimsky points out, the logic of this decision was the recognition of significant modification of natural kinds of living things by human intervention. It is significant that a special statute was required in this case to extend the patent process to specific living entities. Another such act was passed in 1970 to allow the patenting of new hybrid strains of sexually reproduced plants. This time, seeds were included, with profound consequences.

Prior to a crucial 1980 Supreme Court decision, 'no-one could claim monopoly control over the organism independently of how it was used in an invention', without a special statutory intervention. The crucial case in question, *Diamond v Chakrabarty*, involved a claim to patent both a process for breaking down oil from oil spills using a soil micro-organism called *Pseudomonas*, which had been modified through mixing of plasmids (small ring structures of DNA that supplement bacterial chromosomes) from different strains, and for the new strain itself. The claim was that the organism now contained two different oil degrading mechanisms, not previously found together.

Krimsky explains that the application referred to three different sorts of claims, relating to methods of production of the bacteria, the material to carry them and the bacterium itself. The patent examiner allowed the first two claims while rejecting the third, on the grounds of the bacterium being a living product of nature, and hence not patentable. The US Patent and Trademark Office (USPTO) Board of Appeals accepted that the bacterium was, in some sense, a human creation, but still denied the patent on the grounds of its being a living thing.

> The Court of Customs and Patent Appeals reversed the Board's decision and ... the US Supreme Court affirmed the decision of the ... appeals court that a living, human-modified microorganism is patentable subject matter under section 101 of the patent law as a 'manufacture' or 'composition of matter'. (Krimsky, 2004, p. 62)

The majority cited the previous plant patent acts as demonstrating 'no in principle' objection to life patents, while four dissenters argued that Congress had intended that special legislation was required for particular life patents. This 1980 decision 'gave the USPTO the legal mandate to award patents on life forms and on parts of living things' (Krimsky, 2004, p. 64).

As Krimsky points out, 'a bacterium with 5000 innate genes' and one extra 'inserted into its genome' is now regarded as the property of those inserting the gene. It's as if someone putting wet weather tyres on a car or a kind of window washer transferred from a similar car could thereby claim ownership of all cars fitted with these new parts.

After this, numerous patents were awarded for genetically modified animals, plants and bacteria, human cell lines, including human embryonic stem cells and embryos produced without sperm, antigens, antibodies, vectors, vaccines, and proteins, along with genes and genetic tests. Gene patents are granted to those who locate the alleged gene and identify the nucleotide coding.

> A patent on the DNA sequence as a composition of matter gives the patent holder a right to exclude others from using the sequence for any commercial purpose. (Krimsky, 2004, p. 65)

As Robinson (2005) explains, generally patents have been granted to companies who have sequenced the genes and then converted them to cDNA – complementary DNA, produced as a molecular complement from a single strand of DNA through the use of an enzyme – with the patent granted for the cDNA rather than the gene itself. Patents have also been granted for diagnostic tests that have something to do with the genes in question (pp. 320–1).

We have already touched upon some of the problems with patents in the biomedical area (Chapter 3), and they are considered further in Chapter 9. Their proponents try to justify them as encouraging valuable research. But competitive financial interests now stand in the way of the free flow of information amongst scientific researchers and the public. The complex interactions and interdependencies of genes and parts of genes with other features of the organism and the environment, considered earlier, show how this will inevitably be the case. Indeed, the whole idea of gene patenting seems to be built upon the misunderstandings inherent in the reductionist one-gene, one-protein model.

Not only do such patents obstruct research; they also obstruct effective health care, by increasing costs and denying access to relevant information. Companies have patented genes associated with human illness so they can charge large amounts to test people for the presence of such genes. 'Companies patent disease causing bacteria and viral genes... to block diagnosis and treatment until a hefty licensing fee has been paid' (Robinson, 2005, p. 321).

The future

None of these considerations necessarily imply that genetic engineering has no place in a civilised and sustainable future. Some current technologies of genetic manipulation – involving bacteria – seem less straightforwardly problematic than the technologies which have been the principal focus of concern in this chapter. Insertion of human genes into bacteria can allow the production of proteins with valuable medical applications, of high purity and reduced immune response compared to those obtained elsewhere. The use of monoclonal antibodies, stem cells, and proteomics could well revolutionise medical treatment in a number of areas. There might even be a place for properly developed and tested GE foodstuffs with genuinely valuable new properties in the future.

But as indicated by Posilac, the most extreme care and stringent tests are needed in relation to gene transfer into bacteria. The currently dominant agricultural technologies considered here need to be immediately scrapped. And there are serious legal issues of deception and large-scale genetic pollution of previously non-GM crops.

DISCUSSION TOPICS
1   What should be done about GM crops and animals today?
2   How would a meaningful labelling regime operate?
3   What might usefully be achieved through biotechnology?

Energy and the greenhouse effect

This chapter focuses mainly upon primary energy production, and its ecological and public health consequences, particularly in terms of climate change. Primary energy is all commercial energy obtained directly from the environment, including coal, crude oil, natural gas, firewood and hydroelectricity. Some of these primary energy sources are used directly by consumers, while others are converted into final end-use energy forms (Diesendorf, 2007, pp. 65–6).

Around the world, fossil fuels – particularly coal and gas – are burned to produce electricity and oil is refined to produce petrol, diesel, aviation fuel and other outputs, which are also burned – mainly to drive transport systems. The chemical energy of fossil fuels, coal, oil and gas, is just as much a material foundation of contemporary society and contemporary patterns of health, as the chemical energy of food. Indeed, as we saw in Chapter 5, food production and distribution has increasingly been sustained and expanded through the use of such fossil energy, in the production of agrochemicals and in the fuelling of farm machinery and means of transport.

Electrical energy and fuel oils are used in manufacturing and construction (31% of total final energy end use in Australia), in transport (38%), in mining (6%), commercial and services (7%), and residential and agriculture (3%). In Australia, aluminium smelting uses 13% of total electricity generation (Diesendorf, 2007, p. 67).

We know that oil and gas are running out, threatening a major energy crisis in the near future. We also know that the continued burning of fossil fuels is driving accelerated global warming through release of greenhouse gases, mainly CO_2. Much of the shorter wavelength sunlight falling on the earth is absorbed by the earth and re-emitted as longer wavelength infrared radiation. Greenhouse gases 'absorb some of the infrared radiation that would otherwise escape into space, thus acting like a blanket' around the planet (Diesendorf, 2007, p. 10). Warmer air carries more water vapour which absorbs more heat. And the more sea ice that melts the more heat the sea absorbs to prevent the ice re-forming. Declining oil supplies won't halt greenhouse warming, since we are burning the remaining oil and gas at an accelerated rate, along with ever increasing amounts of coal. It is true that, the more sulphur pollutes the air, the more clouds are produced to reflect the sunlight. But emissions are far outdistancing pollution.

Atmospheric CO_2 has increased from a pre-industrial concentration of 280 parts per million by volume (ppmv) to around 387 ppmv at present, which is an increase of more than 160 billion tonnes from fossil fuel combustion, land clearing and cement manufacture, representing an overall 30% increase (Maslin, 2009, p. 8).

If we include other greenhouse gas emissions of methane (mainly from agriculture and livestock), ozone, water vapour, nitrous oxide and other gases, this increases the atmospheric burden to around 460 ppmv CO_2 equivalent concentration today. This is a greater change than at any time over the last 680 000 years. In the 1990s the growth rate of emissions was less than 1% per year, 'From 2000 to 2005 it was more than 2.5% per year' (Maslin, 2009, p. 14).

CO_2 that remains in the atmosphere continues to produce greenhouse effects for hundreds or thousands of years, while methane decomposes in eight to 12 years. According to Matthews and Caldeira, given existing CO_2 levels, the only way to stabilise global temperature at present day levels would be to reduce CO_2 emissions to zero before 2018 (Spratt and Sutton, 2008, p. 77).

The mean global surface temperature has increased by 0.74 degrees centigrade (dc) (plus or minus 0.05 %) from 1906–2005, mostly in the period from 1950 (0.7 dc). The global sea level has risen by up to 22 cm over the last 100 years, with over 40 mm increase from 1993–2007. There have been significant shifts in 'the seasonality and intensities of precipitation, changing weather patterns, significant retreat of Arctic sea ice and nearly all continental glaciers'. Strong hurricanes and wildfires are more frequent and intense. 'Over the last 150 years, the 12 warmest years have all occurred within the last 13 years' (Maslin, 2009, p. 59). With Arctic ice at historically low levels from 2005–08, ice free summers are predicted for the near future, leading to rapidly accelerating ocean warming, due to loss of reflectivity (Spratt and Sutton,

2008, pp. 17–18). With CO_2 dissolving to produce carbonic acid, the oceans are now 30% more acidic than at the beginning of the industrial revolution. This is already killing algae that sequester carbon, and other marine species are threatened (Maslin, 2009, p. 100). The Greenland ice cap is now melting at 100–250 cubic kilometres per year (Maslin, 2009, p. 33; Hansen, 2009).

Fossil fuel combustion accounts for four fifths of global greenhouse emissions, from energy production, industrial processes and transport, particularly burning coal to produce electricity and oil to power cars. The other fifth comes from land use changes, primarily the cutting down of forests for the purposes of agriculture, urbanisation and roads (Maslin, 2009, p. 33; Hansen, 2009). When large areas of rainforests are cut down, the land turns to less productive grassland and ultimately to desert, with much reduced capacity for CO_2 storage.

Many climate scientists argue that at least a 50–60% reduction on current levels of global greenhouse gas emissions by 2050 and an 80–90% reduction by 2100 are necessary to avoid catastrophic climate change (Diesendorf, 2007, p. xvii). According to Spratt and Sutton (2008, p. 92) a reduction of emissions to 80% less than the 1990 levels by 2050 is needed to stabilise the atmosphere at 450 parts per million (ppm) CO_2 equivalent. James Hansen of the National Aeronautics and Space Administration (NASA) argues that CO_2 levels will need to be reduced from the current level of around 385 ppm to at most 350 ppm to maintain the habitability of the planet. This can only be achieved by 'phasing out all conventional coal burning by 2030' and by 'aggressively removing CO_2' from the atmosphere by massive reforestation and agricultural reform (Flannery, 2009, p. 43).

Michael Raupach of the Commonwealth Scientific and Industrial Research Organisation (CSIRO) and Global Carbon Project, says that carbon emissions in Australia have grown at about twice the global average during the past 25 years, and at about double the emissions' growth rate in the US and Japan. The Rudd Labor Government acknowledges a 1% per year increase since 1995. A World Bank emissions report of 2007 has Australia increasing its CO_2 emissions by 38% between 1994 and 2004, to become the sixth highest per capita emitter. This excludes reference to forest destruction, where Australia leads the developed world. In absolute terms, Australia's increased emissions over this period were greater than the combined increase of emissions from the UK, France and Germany (Spratt and Sutton, 2008, p. 84).

Australia is the world's largest coal exporter, the fourth largest coal producer and the biggest per capita emitter of greenhouse gases in the developed world. Australia currently generates 85% of its electricity from coal and the principal mode of transport for people, as in the US, is the motor car, with trucks as the main means of freight transport. Australia supplied much of China's coal, which led to that country's emissions rising by 7.5% from 2005–06, with China overtaking the US, by an 8% margin, as the world's largest emitter in 2006 (Tickell, 2008, p. 34). Major government infrastructure programs leading up to the current recession – and stimulus programes when the crisis hit – involved instituting quicker and more efficient ways to move the coal out of the country.

As Diesendorf (2007, p. 215) notes, underground coal mining is dangerous and unhealthy, with miners at high risk from respiratory disease, and from underground explosions and fires. Such mining also degrades surrounding land. Coal burning is a major source of air pollution, emitting large amounts of oxides of nitrogen and sulphur, fine particles and aerosols, heavy metals, including mercury, volatile organic compounds, and fluoride. 'Water is diverted from drinking, agricultural and ecological uses' for use in mining and coal washing, and it is polluted in the process. Hansen (2009, p. 177) points out that mountain top removal mining creates toxic sludge ponds and mining waste dumps in valleys, poisoning water supplies and 'causing multiple documented health problems for nearby populations'.

As Diesendorf shows, when burnt in a modern coal fired power station, the chemical energy of coal only produces electricity at 35% thermal efficiency (TE), (TE (%) = useful energy output x 100 divided by energy input (Diesendorf, 2007, p. 63)), with the rest lost as waste heat, compared with 45% of the wind energy passing through the sweep of a wind turbine blade converted to electricity (Diesendorf, 2007, p. 113). Taking account of the efficiencies of energy conversion in a current coal-fired power station, in transmission and distribution and final use (for example, to produce light) overall efficiency can be as low as 3% or worse (Diesendorf, 2007, p. 65).

Motor vehicles emit carcinogenic organic compounds along with carbon monoxide, toxic oxides of nitrogen and sulphur oxides. As Diesendorf points out, traffic noise causes stress, car dependence leads to obesity and half the land area of urban centres is devoted to motor vehicles.

> With run-off from roads and car parks containing oil, grease and heavy metals, motor vehicles pollute creeks, rivers, harbours and beaches. Roads are responsible for many deaths and disabilities from crashes. In 2005 in Australia, 1636 people were killed in 1481 road crashes. (Diesendorf, 2007, p. 186)

'Peak oil' refers to the time when half of the world's readily accessible oil reserves are gone, with less oil left in the ground than has already been pumped out, leading to accelerating decline in output and increase in costs in subsequent years. Some experts think that peak oil has already occurred. As Diesendorf says, in light of the end of cheap oil, 'major investment in the development of urban public transport, railways, organic agriculture and renewable energy are essential' (Diesendorf, 2007, p. 190). In particular, it makes sense to shift to transport based on sustainable electricity production (via bioenergy, wind or solar power). Such steps are conspicuously lacking around the world, particularly in Australia (Diesendorf, 2007, p. 190). Instead, dominant federal transport funding has continued to be spent on roads rather than rail with 'under-recovery of road costs – including deaths and disabilities – from the heavier, long distance trucks' (Diesendorf, 2007, p. 203).

Future CO_2 emissions will depend on economic and population growth, fossil and alternative fuel usage, rates of deforestation and reforestation. The Intergovernmental Panel on Climate Change (IPCC) is a conservative, consensus based organisation employing 400 experts from 120 countries to produce recent reports. It was set up in 1988 by the UN Environment Panel and the World Meteorological Organization to review available information on climate change (Maslin, 2009, p. 14). In 2000 the IPCC provided six scenarios corresponding to a range of CO_2 equivalent values in the atmosphere in 2100 from 600 ppmv to 1500 ppmv, with a global average temperature increase of 1.4–5.8 degrees centigrade (dc) on 1990 figures. 'The IPCC also has a seventh – constant year 2000 – scenario to show what change has already been instigated' (Maslin, 2009, p. 14). Even the constant year 2000 model predicts 0.1 dc per decade increase over the next 20 years or more, with temperatures possibly stabilising at a minimum of 0.6 dc higher than in 2000.

'The best estimates' for the six emission scenarios give global temperature increases of 1.8 dc to 4 dc on 1990 levels and sea level increases of 28–79 cm by 2100. But subsequent emissions increases were actually worse than the worst IPCC prediction in 2000. The IPCC 2007 report came up with a new upper estimate of a 6.4 dc increase by 2100. A number of experts have argued that double pre-industrial levels of CO_2 will mean 6–10 dc temperature increases on pre-industrial levels by 2100.

In 2007 the IPCC was predicting a likely sea level rise of 18–59 cm by 2100 or 2 mm per year, with up to 3 dc of global warming (on 1990 levels). Later that year satellite data showed that actual sea level rise from 1993–2006 was already 3.3 mm per year.

Leading expert on sea level rise, James Hansen, argued that the IPCC estimates were based on predictions of slow and linear melting of the Greenland and West Antarctic ice sheets – leading to 2 cm sea level rise per decade. He argues that available evidence suggests a rapid non-linear melting with a possible doubling of sea level rise per decade with global temperature 2–3 dc higher than today, leading to a 5 m rise by 2095.

In his 2006 report to the UK Government, Nicolas Stern pointed out that '200 million people live on coastal flood plains, with 2 million square kilometres of land and 1 trillion dollars' worth of assets less than 1 m elevation above current sea level. 22 out of the biggest 50 cities in the world are at risk of flooding from coastal surges with a 1 m sea level rise – including Tokyo, Shanghai, Hong Kong, Mumbai, Kolkata, Buenos Aires, St Petersburg, New York, Miami, and London' (Spratt and Sutton, 2008, p. 42).

A significant proportion of the populations of Bangladesh, Egypt, Nigeria and Thailand live by river deltas, with human activity, particularly freshwater extraction through tube and deep wells, causing such coastal land to sink, by up to 2.5 cm per year (Maslin, 2009, p. 84). A World Bank worst case scenario predicts 16% loss of land in Bangladesh by 2100 – land currently supporting 13% of the population and producing 12% of gross domestic product (GDP) (Maslin, 2009, p. 84). As Maslin suggests, a 50 cm sea level rise in the next 100 years could be dealt with by 'the protection and adaptation of coastal regions in most parts of the world' (Maslin, 2009, p. 85). However a 1 m increase would create 'major problems' (Maslin, 2009, p. 86).

Three degrees of warming over a long period are predicted to melt the whole Greenland ice sheet, raising the sea level by 7 m, radically changing coastlines around the world. A 5 m rise would cover more than half of Bangladesh. This could also change the circulation of the water in the North Atlantic, and thereby change the climate of Northern Europe. Weakening of the Gulf Stream will give Europe extreme seasonal weather similar to that of Alaska today. Melting of the West Antarctic ice sheet will produce an 8.5 m rise in sea level.

Computer models show the proportion of rainfall occurring as heavy rainfall will continue to increase, as will the year to year variability. This will increase the frequency of flooding. Summer monsoons will increase in strength with the land heating more than the sea and warmer air holding more water vapour (Maslin, 2009, pp. 86–7).

As global warming increases ocean temperatures there will be more hurricanes, leading to potentially huge loss of life and major economic loss in less developed regions. As Maslin points out, storms and floods have the ability to destroy major cities and food crops. Heatwaves and droughts are also major killers with the 2003 heatwave in Europe killing 35 000 people. Future heightening of El Niño could further intensify monsoons, storms, hurricanes and droughts.

The most important threat to human health is loss of access to fresh water in countries with a high ratio of relative use to available supply and regular drought. Long before rising seas swamp the land, the porous rocks of coastal aquifers that supply hundreds of millions of people with fresh water will be contaminated with salt. This includes the aquifers used by Shanghai, Mumbai, Lagos and Buenos Aires (Spratt and Sutton, 2008, p. 43). The problem will be exacerbated by sinking water tables due to low rainfall in many regions.

As David Archer says

> the greenhouse climate has the potential to produce... 'mega-droughts' lasting a decade or longer... when a drought lasts longer than... a year or two, reserves run out. Extended drought changes the vegetation and soils in ways that tends to 'lock in' the drought conditions. (2009, p. 47)

Changes in river run-off affecting yields from rivers, reservoirs and groundwater supplies, increased evaporation leading to increased salination of irrigated agricultural lands, and saline intrusions into coastal aquifers are predicted to move the world from current levels of 1.7 billion people water-stressed to 5 billion by 2025, according to the IPCC (Maslin, 2009, p. 96).

As noted in Chapter 5, loss of water and land threaten food production. Melting mountain glaciers in the Himalayas supply fresh water and fertilising silt to over a billion people in the Ganges, Indus, Brahmaputra, Meghna, Mekong, Yangtze and Huang He Rivers. Mountain snow holds the winter snowfall, releasing the water in spring and summer when the agricultural need is greatest. Monsoon flooding worsened in the 1990s (Maslin, 2009, p. 84), but future year round melting threatens to reduce such spring and summer water and silt supplies. The glaciers that provide the summer water are also melting in the Peruvian Andes and the Sierras of the American Pacific Northwest (Spratt and Sutton, 2008, p. 97).

Use of food crops and land for bio-fuels, partly resulting from policies aimed to address global warming as well as peak oil, contributed to the food price rises of 2007–08 (with corn prices up by 31%, rice by 74% and wheat 130%) along with increased demand for meat in China and India, and speculation in oil and food. This added millions more to the billion people under increasing food stress. Further food price increases in the future pose a major threat to the life and health of poorer people.

The lesser capacity of developing world farming to adapt to changing climate will lead to greater falls in production levels than in most developed countries. Countries relying on one or two agricultural products for export revenue to import food will be particularly hard hit. Comparative advantage fails to anticipate global warming.

Increased warmth and moisture caused by global warming will enhance the transmission of diseases. Malaria, currently infecting 500 million people, could infect many more with the expansion of areas suitable for transmission if proper public health protections are not in place.

All of these predictions are based upon a linear relationship between greenhouse gas increases and climate change. But there are a number of sources of potentially rapid change with sudden qualitative increases in the warming process. Firstly the release of thousands of gigatonnes of methane, a greenhouse gas that is 21 times stronger than CO_2, from gas hydrates in the permafrost and, more significantly, from below the world's oceans, due to the heating and release of pressure of ice in Greenland and Antarctica, could produce a 'runaway greenhouse' effect, as happened 55 million years ago, with a sudden 5 dc increase as a result of a 1500 gigatonne release. Explosive local releases of gas from hydrates could produce tsunamis throughout the world's oceans, like the Boxing Day tsunami that killed 281 000 people in 2004 (Maslin, 2009, p. 118).

Secondly, as Maslin points out, there is some evidence that the Amazon rainforest is currently delaying global warming by absorbing increasing amounts of CO_2 (possibly an additional 5 tonnes per hectare per year). By 2050, global warming could have increased the winter dry season to the point where forest fires destroy most of the rainforest. Such fires would accelerate global warming and the rainforest would be replaced by dry grassland absorbing much less carbon – causing warming to 'accelerate at an unprecedented rate' (Maslin, 2009, p. 120).

Remembering that 5 dc (above 1990 levels) is no longer the IPCC's worst case scenario, it's worth considering what a 5 dc change would actually be like. As Lynas notes, with 5 degrees of global warming

> The remaining ice sheets are eventually eliminated from both poles. Rainforests have already burned up and disappeared. Rising sea levels have inundated coastal cities and are beginning to penetrate far inland into continental interiors. Humans are herded into shrinking zones of habitability by ... drought and flood. Inland areas see temperatures ten or more degrees higher than now. (Lynas, 2008, p. 193)

In the north, 'all of central America, the southern half of Europe, the western Sahel and Ethiopia, southern India, Indo-China, Korea, Japan and the western Pacific' and in the south, 'portions of Chile and Argentina, eastern Africa and Madagascar, almost the whole of Australia and the Pacific Islands' would become extended desert areas, in the grip of perpetual drought. As Lynas says, 'higher evaporation reduces available soil moisture in semi-arid regions, cutting rainfall further and turning them into full scale deserts' (Lynas, 2008, p. 194).

When considered together with the

> exhaustion of fossil aquifers and the disappearance of snow and glacial melt from mountain chains, these belts imply widening zones of the planetary surface which are no longer suitable for large-scale human habitation. (Lynas, 2008, pp. 194–5)

With the 'belt of habitability' contracting towards the poles, James Lovelock anticipates China invading Siberia and the US invading Canada 'to seize the remaining habitable land by military force' (Lynas, 2008, p. 197). As Lynas points out, 'recently glaciated soils tend to be thin, rocky and poor, with little in the way of nutrients and organic matter ... and summer heat-waves would destroy large areas of boreal forest across subarctic regions of Canada, Alaska, Scandinavia and Russia' and may make the continental interiors too hot to grow crops (Lynas, 2008, p. 196). Such summer time drying would require 'major engineering work such as dams to hold winter rains ... to irrigate any new crops' in the region.

Without highly integrated and large-scale planning, including massive energy use for desalination and irrigation (and probably even with such things), as Lynas says, 'a drastic reduction in human population' resulting from war, natural disasters, starvation and chaos, 'is unambiguously the most likely outcome of a rise in temperature towards 5dc – what Lovelock terms "the cull"' (Lynas, 2008, p. 214). It's difficult to imagine such rational cooperation and planning and equal sharing of burdens at this later stage as the situation get increasingly desperate, if it has not been achieved earlier, when change could have been relatively easy and painless.

Kyoto and carbon trading – the market solution

The *UN Framework Convention on Climate Change* was created at the Rio Earth Summit in 1992 to try to negotiate a worldwide agreement for reducing greenhouse gases. In 1997 the *Kyoto Protocol* was drawn up stating the general principles of a worldwide treaty on cutting emissions with all developed countries aiming to cut their emissions by 5.2% on their 1990 levels by 2008-12. The Treaty was ratified in July 2001. The US withdrew from the negotiations at that time and didn't sign the Treaty. The targets were reduced to a 1–3% cut on 1990 levels to encourage Japan, Canada and Australia to sign. Australia ended up with a commitment to an average 8% increase on 1990 levels for 2008-12.

The Protocol came into force in 2005 with 55 countries, representing more than 55% of world emissions, on board. Since then the EU has turned the Treaty into law calling for cuts of 8% on 1990 levels by 2012. This means that if the targets are attained we will see cuts relative to 1990 levels of between 1 and 8% for fewer than half of the developed world countries. As Maslin says, 'compare this with the scientists' suggestions that up to a 60% cut is required to prevent major climatic change' (Maslin, 2009, p. 133). The biggest world emitters, China and the US, have not even signed. Very few countries are on target and the International Energy Authority predicts massive increases in emissions for the period 2000-30, making Kyoto largely irrelevant.

Tickell notes that

> the rate of increase of global CO_2 emissions from burning fossil fuels and from industry accelerated, from 1.1% per year for 1990–99 to more than 3% per year for 2000–04. (Tickell, 2008, p. 1)

In June 2009, as a prelude to the Copenhagen Summit in December and aiming to renegotiate the Kyoto arrangements, the UN Climate Change Secretariat estimated that then current pledges by the rich countries, excluding the US and Japan, added up to between 6 and 24% below 1990 emission levels in 2020 – far below what is needed to stabilise CO_2 at 450 ppm and avoid a

2 dc temperature rise. With the inclusion of the US and Japan, *Nature* editors estimated a cut of 8–14% by the rich countries by 2020. Factoring in China and India, the Potsdam Institute of Climate Impact Research predicted more than 32% higher emissions in 2020 than in 2000. In the same week, Japan announced an 8% cut on 1990 levels by 2020 and a bill before the US Congress aimed at 4% (Morton, 2009).

In 1997, US delegates pushed strongly for carbon trading as a principal tool of worldwide emissions reduction, without the need for authoritarian and costly direct regulation of industry by government. As Lohmann points out, although the US dropped out of climate negotiations soon after, 'a cluster of world carbon markets [now] constitutes the major international response to global warming' (Lohmann, 2006).

The basic idea here is for government to set a target of allowed emissions and then allocate permits to producers of the emissions- directly or indirectly – while ensuring that the total number of permits is consistent with the target, and then

> mandating that producers of emissions acquire sufficient permits to cover their emissions. Participants . . . may trade the permits . . . in theory, participants who can reduce their emissions at low cost will sell some of their permits to participants who can only reduce their emissions at high cost . . . buying time for those with high cost measures to change their practices. (Diesendorf, 2007, p. 298)

As Diesendorf (2007, pp. 298–9) says, choices have to be made about which gases to include (only CO_2 or others as well), how the relevant emissions are measured, how the permits are allocated (free of charge to emitting industries in proportion to current emissions or auctioned to everyone), and to whom (to major polluters or to consumers).

Lohmann (2006, p. 5) explains the logic of this approach. Say the overall annual cap on a sector's emissions is 100 tonnes, for example, 'the government might require two different industrial installations, A and B, to limit their emissions to 50 tonnes a year each'. With A and B each currently producing 100 tonnes, such a reduction may be very costly for one but much cheaper for the other. If it's cheap for B to reduce emissions to zero but difficult for A to achieve any reduction, carbon trading allows 'B to make A's reductions for A', with A continuing as usual. 'B makes money at the same time as A saves money. Both come out ahead yet the same environmental goal of limiting pollution to 100 tonnes a year is met.' The government further reduces the cap in the future, 'secure in the knowledge' that it is reducing carbon emissions in the 'cheapest way possible'.

The A type industries will be those mainly responsible for global warming, like those producing coal fired electricity, with big investments in long-lived, high pollution operations (Lohmann, 2006, p. 6). These are the industries that need to be assisted – or forced – to immediately start the transition to low emission, sustainable technology or it will simply be too late. However, 'cap and trade' offers a way for them to delay any such 'structural change', through buying pollution permits' (Lohmann, 2006, p. 6). And sectors not covered by the cap can increase their fossil fuel use.

In the absence of reasonably priced low emission alternatives, high polluting oligopoly suppliers, in areas of relatively inelastic demand and forced to pay for permits, can simply pass the costs on to consumers. This highlights the absolute necessity for governments to ensure the availability of such alternatives.

Giving away permits free to high emitters in proportion to their emissions at a certain date encourages such emitters to maximise their pollution prior to that date. It rewards those who have failed to take action in the years since the *Rio Declaration on Environment and Development* of 1992, despite their awareness of the ecological consequences of their actions in the intervening years. Depending on how the rest of the scheme works, new cleaner industries can be forced to buy permits from such older high emitters, suffering economic disadvantage.

High emitters receive this windfall in addition to direct and massive public subsidies received over the years. In Australia, for example, those profiting from production and use of fossil fuels already receive government subsidies of around A$10 billion per year, including the A$200 million per year subsidies the Victorian Government gives to the ALCOA aluminium smelter for cheap coal and electricity. This, of course, doesn't count the actual social costs of their operations.

These are directors and shareholders who have freely chosen to continue business and profits as usual on the basis of such emissions and such subsidies with little or no effort to move to more socially responsible practices. Typically the response of these people since 1992 has been not only to continue to mine, sell and burn fossil fuels on a larger and larger scale, but also, as Hansen (2009, p. 245) points out, to actively promote the continued use of such fuels, through lobbying and financing politicians and funding the production and dissemination of fraudulent 'information' deliberately seeking to 'muddle the issues in the public's mind'.

Prior to 1997, for example, many major fossil fuel sellers and burners, including BP, Exxon Mobil, Ford, and the National Mining and Aluminium Associations, were united in the Global Climate Coalition, lobbying politicians and publicly attacking the credibility of individuals and organisations seeking to communicate the facts about global warming. In Australia, 'captains of [major] polluting industries' boasted that they were responsible for writing the Howard Liberal Government policy on greenhouse response, including slowing the growth of the renewable energy industry (Diesendorf, 2009, p. 24, p. 25).

These people and their political supporters should be subjected to civil action and criminal prosecution, forced to compensate for past harms, rather than being freely rewarded with ownership of the atmosphere and the right to continue to fill it with poisons. In ethical terms, free permits – or any permits – to high emitters is hardly just deserts. Nor is it obvious that such a thing is justifiable in utilitarian terms, given the past record of those involved. On the contrary, justice and logic suggest that rewards should go to those with a past record of low emissions, no emissions or serious efforts to reduce emissions to allow them to rapidly expand their operations.

Supposedly, cap and trade will provide incentives not only for those industries that can rapidly green their operations relatively cheaply, but also for the development of new and cheaper low emissions technology to sell to those industries that cannot. But as Lohmann points out

> smart businesses that attempt to profit from selling carbon pollution rights will concentrate on realizing the cheapest opportunities for emission reductions first, regardless of whether they lead to long term structural change away from fossil fuels. The cap and trade goal of reaching modest numerical emissions targets cheaply is simply not the same as the goal of mitigating global warming, which entails taking immediate steps to break the deeply rooted dependence industrialized societies have on fossil fuels. (2006, p. 7)

Transnational corporate operations can sometimes simply shift their high emissions operations, which are also typically high polluting in other ways, to less developed territories, not covered by the cap. This has been happening for a while. Carbon trading provides a further incentive for differentially exposing increasing numbers of poorer people to toxic emissions.

If most pollution rights are auctioned instead of being given away free then the biggest carbon polluters, along with such speculators as private equity and hedge fund operations hoping to benefit from such polluters, are in the best position to acquire the bulk of such permits, to support continued business as usual.

Current programs of, and future prospects for, cap and trade also involve complementary carbon offsets, which allow 'industries, nations and individuals' to finance 'carbon saving' projects elsewhere instead of cutting their own emissions. 'Examples include tree plantations...hydroelectric dams...efficiency schemes and other projects that "displace" fossil energy' (Lohmann, 2006, p. 10).

The problems here are at least as serious as those associated with the cap and trade system. Like cap and trade itself, such schemes function precisely to allow the rich and powerful fossil fuel burners to continue business as usual. Just as with the original credits, the biggest industrial carbon polluters can buy the bulk of offset credits to sustain continued habitat destruction. The process is open to further horrendous abuse. As Lohmann notes

> the carbon saving of the offset project can only be calculated by showing how much less greenhouse gas is entering the atmosphere as a result of its presence than would have been the case otherwise. That entails identifying a single, unique business-as-usual storyline to contrast with the storyline that contains the project. (2006, p. 14)

Low emissions projects, which undoubtedly would have been instituted anyway (because of lower cost), can be presented as responses to the offset scheme requiring appropriate compensation. Those that pay can increase their own pollution. Business operations in countries without caps can say that they were about to embark on all sorts of horrendous projects, or can actually institute such projects and then claim money for not starting, or for mildly ameliorating, the emissions from the new high emission projects. If the payments are high enough they could effectively subsidise such high emissions projects. Then they can use the extra money to finance yet more high emissions operations. Here, carbon offsets accelerate rather than reduce global warming.

The EU's Emission Trading System (ETS) is the principal means by which the EU 'aims to meet the 8% emissions reduction allocated to it in the Kyoto Protocol' (Tickell, 2008, p. 49). As Tickell points out, there are many issues and problems associated with the scheme. The first phase of the program included only the largest emitters, ignoring 60% of EU emissions, and permits were given to the biggest polluters in proportion to their historic emissions. In other words lack of virtue, rather than virtue, was rewarded. Under pressure from industry, permits were allocated for more emissions than were actually being produced at the time (e.g. 65 megatons (Mt) CO_2 more than in 2005)

> This both reduced the incentive for technological innovation and handed electricity generators as much as 30 billion euros per year in surplus profits as they traded their surplus allowances, and passed the price of their allowances (EUAs) on to electricity consumers, although they were received at no cost. (Tickell, 2008, p. 49)

Needless to say, there were no reductions in emissions and the price of credits fell precipitously. Again, rapidly rising emissions from aviation and shipping were not covered.

Phase 2 (2008–12) involves a 7% reduction in the cap, with medium-sized emitters covered, and more auctioning of allocations. Tickell is confident of significant improvements in Phase 3 of the scheme, with 100% auctioning of credits to utilities (but free handouts preserved for cement, steel and aluminium makers) and overall allocation down 21% from 2005 levels by 2020.

In terms of carbon offsets, as Tickell points out, Kyoto allows businesses in the industrialised countries to compensate for failures to keep within their quotas by

> buying in reductions in greenhouse gas emissions from other countries... These mechanisms have created a dynamic and for some highly profitable business sector... however, in some cases... the emissions reductions are entirely notional and in the worst cases greenhouse gas emissions are actually being stimulated by the Kyoto Protocol, rather than reduced. (Tickell, 2008, p. 35)

Projects that reduce emissions in developing countries can generate credits for sale to developed countries failing to meet their targets. However, there is no proper oversight of such projects. Hydroelectric projects already under way, completed or implicated in human rights abuses and environmental destruction, have been given such credits, as have industrial projects involving massive combustion of fossil fuels and pollution (on the grounds of some 'waste heat' recovery).

It costs relatively little to cut emissions of the greenhouse gas HFC-23 from processes producing refrigerant gases. Developing world producers are now receiving twice as much from carbon credits as they can get from selling refrigerant gases – to the tune of £4.7 billion by 2012. This money goes to big industrial polluters who then use it to expand their polluting operations (Tickell, 2008, pp. 39–40). As Lohmann points out

> The Indian company SRF plans to take a US$600 million profit from selling UN... licenses to Western companies and invest it in a new plant producing a gas 13 000 times more... damaging than carbon dioxide, HFC 134a. (2006, p. 14)

The US$600 million was gained by spending US$1.4 million on reducing the output of a gas called HSC-23 from an existing factory. Purchasers of these credits included 'Shell International Trading, Barclays Capital and Icecap, a London based emissions trading company' (Lohmann, 2006, p. 14). SRF's operations have been criticised by local residents claiming pollution of crops and water supplies.

As Lohmann says

> while the biggest northern buyers of carbon credits include such large scale corporate greenhouse gas producers as Shell, BHP Billiton, EDF, Endesa, Mitsubishi, Cargill, Nippon Steel, ABNJ Amro, and Chevron... major sellers comprise... strikingly similar corporations in the south [including] the Tata Group, ITC, Birla, Reliance and Jindal [in] India, Korea's Hu-Chems Fine Chemical, Brazil's Votarantim and South Africa's Mondi and Sasol. (2006, p. 13)

As Shiva (2008, pp. 22–3) observes, rather than the credits going to Indian tribal farmers providing fresh vegetables with minimal use of water and fossil fuels, they go to sponge iron producers using lots of water and fossil fuels, poisoning such farmers' crops, air, soil and water on a large scale, because sponge iron production involves less direct CO_2 emissions than blast furnace production. Clean Development Mechanism (CDM) projects in India do not support ecological farming or sustainable pollution free technologies. 'CDM projects do include coal fired power plants and automobile factories... petrochemical plants and chemical factories, cement industry and sugar factories...'

While deforestation, particularly in Brazil and Indonesia, now accounts for 18% of global emissions, only a single forest project has been approved for carbon credits. As Tickell (2008, p. 42) says 'the conservation of existing forests is not accommodated within the CDM [credit system], only the establishment of new forests'. Millions of hectares can be destroyed at no cost with a few hectares replanted for money from the EU.

Kyoto applies no meaningful sanctions to those failing to meet targets, sets no targets for developing countries, provides inadequate funding for adaptation and omits 'fast-rising emissions from shipping and aviation' (Tickell, 2008, p. 41). It lacks any clear scientific or economic basis and has radically failed to 'constrain the overall rise of global greenhouse gas emissions' (Tickell, 2008, p. 43).

The UN Climate Change Conference in Bali, 2007 produced plans for further discussions on the subject of reducing emissions from forest destruction, by extending the carbon trading regime. As Tickell notes

> we need to cut industrial greenhouse gas emissions and to save the world's forests, not one or the other. By putting carbon credits from reduced emissions from forest degradation and destruction into the... carbon trading regime, we are accepting that the Annex 1 countries will be able to continue to pollute at will, providing they offset their pollution by reducing emissions from deforestation elsewhere. (2008, p. 46)

Destruction of bio-diverse natural forests to allow for 'reforestation' by (non-native) eucalyptus monocultures could well earn credits under the proposed new system. The EU's Phase 3 ETS sets a cap on the volume of EU credits that can be substituted for Kyoto credits, and will exclude forestry-based credits altogether, forcing the majority of emissions reductions to be achieved within Europe' (Tickell, 2008, p. 51).

Australia

The Australian Labor Party's proposed Carbon Pollution Reduction Scheme is a particularly dismal effort, especially considering Australia's particular culpability as coal exporter and emitter. The cap was set at 5% below 2000 levels by 2020, with a guarantee of five years' notice of cap changes to industry.

The Labor leader, Kevin Rudd, said that the cap would be increased to 15% or possibly even 25% if others agreed to similar levels at the Copenhagen Summit. The scheme includes 1000 companies in the energy, transport, manufacturing and oil and gas sectors, and, supposedly, 75% of emissions of all six major greenhouses gases. However, it doesn't include agriculture and there is no special body to police it, with corporations reporting their own emission levels at the end of each year and surrendering a permit for each tonne of emissions produced in that year. The permits are distributed by a combination of administrative allocation and auction, with subsequent secondary market trading between businesses. The permits also have no expiry date. The government says that the price of carbon will be determined 'by supply and demand' for permits in such auctions and secondary markets. But the government has set a price cap of $40 per tonne for five years and anticipates a market price of around $25. Companies are allowed to buy unlimited Kyoto overseas carbon offsets.

A 5% emissions reduction on 2000 levels by 2020 is completely out of kilter with the Labor Party's own *Garnaut Climate Change Review*, which argued for a 25% reduction on 2000 levels by 2020 (and 90% by 2050) or a minimum of 10%, while acknowledging that the latter level of reduction was not compatible with 'a good world'. As Tony Kevin points out, a 5% level is so low that it will be achieved by voluntary community emissions savings, with big electricity generators 'under no economic pressure to change their present technologies at all' and left with unused credits to sell to other polluters, or offer more cheap electricity to 'big polluters like aluminium plants' (Kevin, 2009, p. 9, p. 24).

As Diesendorf (2009, p. 54) notes, '$25 per tonne would not enable renewable energy to compete with coal or gas'. He goes on to say, 'separate policy measures are crucial to promote rapid development and deployment' of emission free technologies right now. People can only choose to buy clean power if someone is offering it at an affordable price. Permits as 'permanent property rights' rather than temporary licences with expiry dates raise the possibility of tax payers having to pay for huge amounts for future government buy-backs as the true reality of the situation becomes increasingly apparent.

The original proposal in December 2008 gave what were called emission intensive, trade exposed exporters 90% of their permits free, while other sectors received 60% of their permits free. This was later increased to 95% and 66% respectively and justified as necessary to avoid 'carbon leakage', with companies relocating overseas. But such industries are actually identified by reference to emission levels per million dollars of revenue. Such assistance will be reduced by 1.3% per year. A percentage of electricity consumed by a company also counts towards their carbon emissions.

Any business that increases its output would receive more free permits in proportion to the increase and 'new investments' would receive assistance at the same rate. At the same time, any reduction in demand for high emission goods, leading to a reduction in output by the industry concerned, would result in no reduction in free permits. So that reduced demand for coal produced electricity due to increased use of solar panels would leave the electricity industry free to sell excess permits to allow increased pollution elsewhere.

In May 2009, the government announced a scheme to allow households installing emissions cutting technology to buy carbon credits to remove such credits from circulation. As Kevin (2009, p. 9) says, 'it is hard to imagine' and totally unreasonable to expect that people should thus pay twice for reductions and for permits.

Nearly half of the money raised from sale of permits is supposed to go to poorer households as cash handouts to compensate for price rises by the polluters. Coal fired electricity generation is identified as the most 'strongly affected industry' despite not being 'trade exposed', in part because producers 'cannot pass on cash costs' if they are competing with sustainable energy! They are given a 'once and for all' administrative allocation of permits to the value of $3.9 billion based on a carbon price of $25 per tonne over the first five years of the scheme (Carbon Pollution Reduction Scheme, 2009, http://www.climatechange.gov.au/publications/cprs/white-paper/cprs-whitepaper.aspx).

Kevin Rudd claims that the plan is consistent with CO_2 levels of 550 ppm by 2020, if other nations go down a similar path. His party's own *Garnaut Climate Change Review* argued that a 10% reduction was necessary to achieve 550 ppm. The Australian National University's Centre for Climate Law and Policy says the figure for Rudd's reduction is actually 650 ppm.

As the Australian Government carbon-trading White Paper says, carbon trading schemes are already operating in 27 EU member states, with 27 states and provinces in the US and Canada introducing such schemes, and a New Zealand scheme in operation since 2008. But as Kenneth Davidson of *The Age* newspaper points out, no such schemes have as yet resulted in any reduced emissions from the pollutants subject to the caps, while the Scandinavian countries along with Holland and Italy, which have instituted carbon taxes, have achieved some reductions. Davidson calls for an immediate Australian carbon tax of $10 per tonne to provide $1 billion per year for renewable energy development – with the Mandatory Renewable Energy Target (MRET) doubled to 40% by 2020, and a 'crash program' of building base-load renewable energy generators (Windisch, 2009).

Responsible governments clearly need to set strong targets for greenhouse gas reductions and use of energy efficient and renewable electricity. They need to immediately halt all construction of coal-fired power stations, hasten the closure of existing ones and end all subsidies to fossil fuels. They need to build appropriate infrastructure to support renewable energy, including power lines for large-scale wind and solar power distribution.

Before its election the Labor Government promised $650 million for the development and deployment of renewable energy. None of this was allocated in its first budget in May 2008. The government also delayed expanding the MRET, as it had promised. However $500 million was given for a National Clean Coal Fund.

In his May 2009 budget, Kevin Rudd committed $1.5 billion for the construction of up to four solar power stations (over a six year period), but $2 billion for 'clean coal' research (over nine years), and $4.3 billion as the first stage in a promised total of $43 billion for a new national broadband network infrastructure.

Despite pumping billions of dollars into the economy to counteract the deflationary effects of the meltdown, the government has completely failed to begin to provide the infrastructure for a low emission economy, including the necessary transmission lines for large-scale wind power systems – particularly in South Australia – and massive reforestation to restore soil fertility and provide renewable carbon neutral bio-fuel for gas turbines, to produce both peak and base load electrical power.

Energy saving, wind and solar power

Huge energy savings and greenhouse gas reductions are possible through effective insulation and draught exclusion from homes and offices, use of solar water heating systems, passive solar heating of buildings, electric heat pumps, energy efficient refrigeration and lighting. New buildings can be equipped with boreholes underneath them for ground sourced heat pumps. The ground warms cold water pumped into these boreholes, reducing hot water costs. More efficient supermarket refrigeration and lighting could produce significant savings, and more efficient kilns, boilers and electric motors for industrial operations could also contribute (Diesendorf, 2007, p. 87).

Dry granite rocks several kilometres underground, heated by the decay of radioactive isotopes, can be used for electricity generation by pumping cold water down to them and using the hot water returning out of another hole. The drilling at first uses existing fossil fuels and a closed cycle is necessary to avoid radioactive contamination. Such geothermal power could be produced at many places around the world, including parts of South Australia and Queensland, and could make a significant contribution to cutting greenhouse emissions in the future. The Geodynamics Company was generating geothermal electricity in the Cooper Basin of northern South Australia by 2009 (Flannery, 2009, pp. 58–9). The project then ran into problems and had to be temporarily closed down, but could easily be restarted.

A number of different wave power technologies also look promising for the future, including conventional tidal power stations, turbines located underwater in powerful ocean currents and deep ocean wave power systems. There are major ecological problems with conventional tidal power and most authorities agree that quite a lot more work is needed to improve on current wave power systems.

There are major problems with hydroenergy projects – they destroy fertile land behind the dam and prevent nutrient rich silts and water getting downstream, while the rotting vegetation in the flooded area gives off significant amounts of methane. However, there are existing technologies which can effectively address global warming around the world.

Since 2003 Denmark has generated 20% of its electricity from wind power, allowing some coal fired power stations to be retired. One of Denmark's wind energy companies has just embarked on a project to create a nationwide recharging system for electric cars, driven by surplus wind power, supported by a tax-free scheme for such vehicles (Flannery, 2009, pp. 66–7). There have been no major problems from wind variability in Denmark. Published wind energy surveys suggest that 20% of Australia's electricity could also be derived from wind power (Diesendorf, 2007, p. 123). 'The 2005 Report *Wind Force 12*, demonstrated that there were – at that stage – no technical, economic or resource barriers to supplying 12% of a – two thirds greater – world electricity demand by 2020' (Diesendorf, 2007, p. 126).

'Wind energy at appropriate sites is the most economical of all renewable energy sources other than large scale hydro-electricity' (Diesendorf, 2007, p. 124). If even a small percentage of 'real costs' of greenhouse gas emissions and air pollution were factored into the monetary cost of electrical energy, wind energy would already be very much cheaper than coal energy. Due to the intermittency of wind, 'it is not possible to generate 100% of grid electricity from the wind' alone without long-term storage 'in the form of hydrogen or advanced batteries' not yet available (Diesendorf, 2007, p. 122). But the larger the scale of wind power operations, the less intermittency is an issue because of a lack of correlation between wind speeds over larger areas (Diesendorf, 2007, p. 119). As Diesendorf says

> a mix of different kinds of renewable energy sources with different kinds of variability – such as wind, solar, wave and bio-energy – could provide a complete generating system. (2007, p. 122)

Solar electricity production similarly involves no emission of greenhouse gas, noise or chemicals. There are few moving parts and 'the energy inputs to, and emissions from, the construction of solar equipment are small and decreasing'. With solar thermal electricity (STE), mirrors of different kinds (including parabolic troughs, dishes and arrays of flat mirrors on towers) are used to channel solar energy which heats up liquid to drive turbines and create electricity. The best places for such solar heat plants are in low latitude deserts with few cloudy days. Such plants have operated in California since the 1980s. Solar panels (PVs), using sunlight to dislodge electrons to create a current, are currently only 17% efficient and require expensive silicon. They can be put on all rooftops and are reliable and maintenance free. Significant improvements are in the pipeline. Wind and solar generation can effectively complement each other on a large scale, with the former more effective at higher latitudes (Diesendorf, 2007, p. 122).

Bio-fuels

Bio-fuels were mentioned in Chapter 5 because of their role in diverting food crops away from the stomachs of the poor to the petrol tanks of the rich. They have contributed to rising food costs which further deprived the poor of food (with the US ethanol program said to have been a major contributor to maize price increases of 60% from 2005–07), and to accelerated global warming, reduction of biodiversity and soil erosion via use of fertilisers and the destruction of tropical rainforests and peat land to make way for palm and soy plantations (for biodiesel).

Oliver Tickell highlights the particular problems of the swamp forests of Indonesia

> which are being logged, cleared, burnt and drained to make way for vast palm oil plantations. The biofuel policies of the EU and the USA are thus accelerating the extinction of the orang-utan and the destruction of the Asian peat-lands... containing 42 gigatonnes of embodied carbon [representing 155Gt of CO_2]. This peat-land carbon is currently being released at 2Gt per year [as drainage leads to peat decomposition] – equal to the entire savings the Kyoto Protocol is expected to deliver during 2008–12. (2008, p. 4)

Burning bio-fuels produces carbon dioxide, just like fossil fuels, thereby contributing to global warming. Production of bio-ethanol and biodiesel from corn, wheat or barley requires high energy input generating more greenhouse gases. An increasing number of concerned people have therefore called for an end to the production and use of such fuels.

However, unlike fossil fuels, biomass can be regenerated on an ongoing basis. As the crops grow, they remove CO_2 from the atmosphere to form carbohydrates and oxygen. When biomass is converted into energy it emits the CO_2 it originally absorbed. In theory, the use of biomass energy could therefore be neutral in terms of CO_2 emission, provided an equivalent amount of biomass is regrown. If CO_2 neutral bio-energy is used as a substitute for fossil fuels, it can reduce CO_2 emissions and air pollution. Biological materials can be substituted for other greenhouse intensive materials like cement. Bio-fuel does not have to be produced at the expense of food crops. Its production can increase food crop production by restoring the fertility of degraded land and providing substantial energy that can be transformed into increased fertility in other ways.

As Diesendorf argues, there is huge potential for sustainable use of bioenergy in Australia, provided the biomass is grown on land cleared prior to 1990 and used at a rate no faster than it can regrow.

> It can enhance energy security by substituting for a significant fraction of imported oil . . . and has potential value for adding to animal habitat and biodiversity. (2007, p. 131)

As noted, there are clearly major problems when food crops are used to produce fuel or dedicated fuel crops take fertile land away from food production. This is happening on an increasing scale in the US with corn for ethanol, in Europe with rapeseed for biodiesel, and in Brazil with sugar cane for ethanol. Furthermore, the production of steam to convert bio-solids to liquid fuels using coal produces lots of extra greenhouse gases and the overall energy inputs to such bio-fuel production (of fertilisers, tractor fuel, fermentation, distillation and plant) have been found to involve only a very small (if any) positive net energy gain from burning the fuels.

But solid biomass can itself be used to fuel the liquification process, and increased energy efficiency can be achieved making ethanol from wood. We must remember that, as yet, it is still purely economic, rather than biological, constraints that are contributing to world hunger.

In the past, Australia has mostly used solid residues from food, sugar and fibre crops to produce bioenergy, mainly bagasse – a fibrous residue of the sugar industry – burned in power stations (along with coal) to generate electricity. As long as sugar production continues there will be substantial amounts of bagasse available to allow for a 'generating capacity of 1650 MW electrical and annual energy generation of 9.8 terawatt hours or 3PPJ' (Diesendorf, 2007, p. 139). Stubble residues from grain crops, which offer little benefit to the soil and its organic content, offer another free fuel for future electricity production.

Such solid residues can be turned into wood pellets and liquid and gaseous bio-fuel that can cost much less to transport and burn in more efficient and less polluting fashion. They can be used in electricity generation, transport and home heating. 'The combustion of ethanol has reduced emissions of sulphur, most organic compounds, particularly CO and hydrocarbons' (Diesendorf, 2007, p. 139). Like burnt methanol, ethanol emits increased carcinogenic formaldehyde. Butanol and biodiesel seem to be much safer. Bio-oils produced from wood are 'possible substitutes for heavy and light fuel oils – including diesel' (Diesendorf, 2007, p. 136).

By far the greatest potential in Australia is that associated with planting large numbers of trees and other deep rooted perennial plants on

> agricultural and pastoral lands ... damaged by dry-land salinity, acidification, chemical pollution, nutrient depletion, compaction and loss of biodiversity. (Diesendorf, 2007, p. 141)

As Diesendorf says, not only is this a huge potential source of substantial bio-energy, but it also allows for 'slowing and possibly reversing' most of these major problems and thereby further combating global warming, extending biodiversity, maintaining agricultural production and creating many new jobs.

In Chapter 5 we considered the major problems of salination and water-logging in Western Australia (WA) as a result of clearing native vegetation to make way for industrial agriculture. We also considered the vital need for replanting native trees to try to save this land. As Diesendorf points out, planting mallee eucalypts 'in belts along contours over 15% of the wheat land where they intercept the downward movement of the water that is surplus in the annual plant agriculture' could allow for harvesting 15.4 million hectares (m ha) of the wheat belt farmland. This could produce 35.7 gigawatt hours (GW-h) of electricity, 3450 tonnes of activated carbon and 1050 tonnes of eucalyptus oil (Diesendorf, 2007, p. 139).

Australia wide, with no displacement of any food crops, mallee planting at this level in degraded farmland could produce 25–30 terawatt hours (TW-h) of electricity. Not only would this not displace food crops, but it could be the only way to maintain such food cropping. Diesendorf argues that much higher density planting over much larger areas (extending beyond the wheat plantings) could produce 67–80 TW-h.

Throughout the country the potential exists for much larger plantings of bio-energy crops to address issues of soil acidification (up to 43 m ha by 2050), soil decline and loss (up to 10 m ha), river degradation and loss of biodiversity, as well as dry land salinity (up to 15 m ha), while supplying a large percentage of the country's demand for oil (mainly ethanol and methanol) and electricity (from 40% efficient integrated gasification combined cycle power plants) by 2050 (Diesendorf, 2007, pp. 146–7).

Carbon produced through pyrolosis of crop residues and regular harvesting of new eucalypt forests can be put back into the soil to increase fertility and sequester carbon, through encouraging the growth of bacteria and fungi 'essential to healthy plant growth' and retaining water and nutrients (Flannery, 2009, p. 83).

Whereas carbon in trees is highly volatile, 'in the sense that if the tree rots or burns the carbon will quickly be released back into the atmosphere', charcoal is relatively inert, 'ploughed back into the soil, charcoal won't readily rot or burn' (Flannery, 2009, p. 81). And some estimates suggest the possibility of 'pulling nine gigatonnes of carbon per annum out of the atmosphere' with soil sequestration of charcoal (Flannery, 2009, p. 84).

CSIRO scientist Barney Foran highlights the need to combine pyrolosis and carbon burial with large-scale tree planting, in order to 'save Australia's farming communities'. He says

> In 2050, an Australia with 40–60 million hectares under farmed forests and wood crops could be self-sufficient in liquid fuels . . . Regional processing hubs could make fuel, bio-electricity, bio-char for carbon storage and soil improvement and green chemicals . . . (Flannery, 2009, p. 170)

Clean coal and nuclear energy

While some nations are winding back the use of coal in favour of gas and other energy sources, coal still remains the principal foundation of electricity generation around the world. Wealthy, powerful and politically influential producers are deeply involved in long-term investment in coal fired electricity generation. These people, and their friends in politics, talk a lot about 'clean coal' and 'carbon capture' as ways of reducing emissions while continuing to burn coal. In New South Wales (NSW) and Queensland, state governments themselves own and run coal fired power stations, and lead the way with the clean coal rhetoric.

As Diesendorf says

> the only genuinely 'clean coal' would be one that is converted to a clean fuel, such [as] hydrogen, by using renewable energy sources. But then, if we had low cost renewable energy technologies, why would one bother to use them to convert coal? It would be simpler, cleaner and probably less expensive to use renewable energy to produce hydrogen directly from water. (2007, p. 225)

He shows that there are currently four main potential 'methods of capture of CO_2 from large point sources' such as power stations and industrial plants (Diesendorf, 2007, p. 229). Carbon dioxide in natural gas fields can be 'pumped down into a separate aquifer'; the flue gas from a coal fired power station can be 'scrubbed' with an amine solution then reheated to release high purity CO_2; coal can be turned into gas through heating with steam and oxygen, with the gas then burnt in a gas turbine with the reduced CO_2 produced captured in similar fashion; and the coal can be burned in oxygen rather than air, followed by removal of CO_2 from the resultant gas by cooling and compression (Diesendorf, 2007, pp. 229–30).

'Once captured, CO_2 can be compressed and transported in high pressure pipelines to the sequestration points' (Diesendorf, 2007, pp. 229–30). Here again, there are four main types of sequestration under investigation: injection into underground geological formations such as depleted oil wells and mine seams; capture by vegetation and soils; injection into the deep ocean; and conversion into another solid chemical compound with reuse (Diesendorf, 2007, pp. 230–1).

As Diesendorf (2007, p. 245) acknowledges, geo-sequestration 'could possibly begin to make a significant contribution to greenhouse gas abatement after 2030'. This is likely to be much more costly, complex, and wasteful than 'a mix of efficient energy use, natural gas and lower cost renewable energy sources such as wind power and bio-energy from crop residues'. It will also be potentially much more dangerous – insofar as the escape of sequestered carbon is capable of poisoning large populations and producing rapid global warming – and continued coal production involves 'air and water pollution, large water use, land degradation, biodiversity loss from coalmines and occupational health and safety hazards'.

As Maslin (2009, p. 160) says 'removal and storage costs of CO_2 could cause a 100% increase in power production costs'. However, 'the major problem with all of [the proposed] methods of storage is safety. CO_2 is a very dangerous gas because it heavier than air and can cause suffocation'. He refers to the sudden release of CO_2 from Lake Nyos in Cameroon in 1986, which killed people and livestock 25 km away, to show that 'storage of CO_2 is very difficult and potentially lethal'.

Nonetheless

> the possibility of geo-sequestration in the future is being used to divert funding away from cleaner technologies that are more cost-effective now . . . and to deflect public attention away from current proposals to build many more conventional coal-fired power stations with high greenhouse gas emission intensities. (Diesendorf, 2007, pp. 245–6)

Presumably the possibility of 'clean coal' also functions to combat criticism of Australia's continued reliance on coal exports.

Tim Flannery (2009, p. 52, p. 45) acknowledges that clean coal technology 'may prove as dangerous as nuclear power and as expensive as solar panels'. He argues that China has 'gone so far down the road of using coal' with no chance of its 'simply knocking down its newly constructed power plants' that 'carbon capture will have to be retrofitted to these plants and ways found to cover the costs'. But it would seem to make more sense, and probably cost less in the longer term, to find a way to cover the costs of knocking these power stations down and building windmills.

Australia is also a major exporter of uranium, and while this has, in the past, divided the Labor Party, the previous, staunchly neoliberal Liberal Government was a supporter of nuclear power at home and abroad as a significant 'solution' to global warming. The current Labor Environment Minister, Peter Garrett, who was previously an anti-uranium mining activist and environmentalist, approved the extension of the existing Beverley (uranium) Mine in

2008 and, more recently, the opening of the new Four Mile Mine in northern South Australia.

Nuclear power stations use the heat generated by splitting of heavy uranium-235 or plutonium-239 nuclei into lighter nuclei to boil water to produce steam to drive a steam turbine to turn a generator. Like the coal industry, the nuclear industry also has powerful political friends. Not only John Howard in Australia, but also the equally committed neoliberal George Bush in the US and Tony Blair in the UK, were all strongly pro-nuclear. Since 2000 the nuclear industry has run a substantial international media and lobbying campaign a promoting nuclear energy as solution to global warming (Diesendorf, 2007, p. 247). Environmentalist James Lovelock has supported the technology as the only one capable of producing enough CO_2 free energy 'in time' to stop catastrophic global warming (Diesendorf, 2007, p. 156).

On the other hand, Diesendorf points out that there are substantial CO_2 emissions from the nuclear fuel chain, including mining and milling the uranium, fabrication of the fuel, enrichment of the fuel to increase the concentration of U-235 from 0.7 to 3.5%, construction of the power station, reprocessing of the spent fuel, waste management, and decommissioning of the power station. Only the actual operation of the power station is carbon emission free. However, such stations periodically 'vent' radioactive gases out into the environment.

As Caldicott says

> abnormal releases of large quantities of radiation at nuclear power plants occur not infrequently and are referred to by the . . . industry as 'incidents'. These 'incidents' occur because of human or mechanical error or because the operator at the reactor has purposefully decided to vent radioactive gases to get rid of them. (2006, p. 64)

Emissions include isotopes of noble gases, which accumulate in animal fat and irradiate surrounding tissues, iodine-131 and strontium-90, concentrated in milk, the former absorbed by the thyroid gland, the latter into bones and teeth, from where they too continue to irradiate surrounding tissues, causing thyroid cancer, bone cancer and leukaemia. Caesium-137 concentrates in muscles and caused an epidemic of a rare cancer called rhabdomyosarcoma in children living near an old reactor in Long Island in the 1980s (Caldicott, 2006, p. 64).

Tritium, a radioactive isotope of hydrogen, 'is released continuously from reactors into the air and into lakes, rivers or seas, depending on the reactor location' (Caldicott, 2006, p. 56). It has a comparatively short half-life of 12.4 years, but this still leaves it radioactive for 248 years. As Caldicott points out, tritiated water is readily absorbed through the skin and is highly mutagenic and carcinogenic. It can also become organically bound to molecules in food, from where it becomes organically bound in the human body, irradiating tissues for years on end.

The Chernobyl meltdown released a half a tonne of plutonium, theoretically enough, as Caldicott (2006, p. 62) points out, 'to kill everyone on earth with lung cancer 1100 times, if it were to be uniformly distributed into the lungs of every human being'. Plutonium is transported from the lungs to the lymph glands where it causes lymphoma or leukaemia and from there into the blood where it causes bone cancer and leukaemia. Also 'it is stored in the liver where it causes liver cancer, and it is teratogenic, crossing the placenta into the developing embryo' (Caldicott, 2006, p. 61). It is also stored in the testicles, where it causes cancer and mutations of the reproductive genes. 'The half life of plutonium 239 is 24 400 years, so it remains radioactive for half a million years' and every male in the northern hemisphere has a tiny amount of plutonium in his testicles from fallout, still falling on the earth from weapons tests in the 1950s and 1960s' (Caldicott, 2006, p. 61).

As Caldicott points out, the burning reactor at Chernobyl spread 400 Hiroshima bombs' worth of radiation across Europe over a period of 10 days. She quotes a 2007 article predicting 800 000 cases of childhood leukaemia in Belarus and 380 000 in Ukraine as a result of fallout, along with UNICEF reports of over 40% more nervous system, bone and muscle, and cardiovascular disorders in the region from 1990 to 1994. The Ukrainian Health Ministry reported massive increases in miscarriages, stillbirths and deformities. 'The WHO predicts that 50 000 children in Belarus will develop thyroid cancer during their lifetime' because of exposure to radioactive iodine following the accident (Caldicott, 2006, pp. 116–17). She refers to estimates of between 5000 and 10 000 of those heavily exposed through participation in clean-up operations as already dead from radiation induced illnesses and '5 million people' still living in 'contaminated areas 5 years after the accident' (Caldicott, 2006, p. 118). At the time of the accident, the US Nuclear Regulatory Commission estimated a 45% chance of a larger meltdown in the US over the next 20 years.

Uranium mining can be lethal to miners exposed to radioactive gas, and to others around the mine sites. Shiva (2008, p. 28) points out that the Jaduguda Uranium Mine in India 'impacts the 30 000 people living in 15 villages within 5 km radius of the complex' (Shiva, 2008, p. 25). Children born close to the tailing dam have 'deformities, skeletal distortions, partly deformed skulls and organs... more than 7000 mine workers are continuously exposed to radiation hazards' (Shiva, 2008, p. 26).

Uranium mining produces lots of long-lived low-level radioactive waste that is kept on site but leaks out into water supplies. Highly radioactive spent fuel is stored for decades under water to allow shorter-lived elements to decay and the temperature of the spent fuel to drop. It was originally intended to extract unused and new fuel for future use. However, such material proved highly dangerous and difficult to handle with major leakages leading to the closing down of the UK reprocessing plant at Sellafield. 'In practice, very little plutonium is being recovered and vast quantities of nuclear waste containing highly radioactive [materials] are in temporary storage at nuclear power stations' and other facilities (Diesendorf, 2007, p. 250). A terrorist strike on one such facility could render a whole US state or the whole of the UK uninhabitable.

No way has been found to safely store or dispose of existing nuclear waste which will be around for thousands of years, let alone the wastes of a future world economy based on nuclear power. No full sized nuclear power station has ever been properly decommissioned. 'Costs could be as high as the original capital cost of the power station' (Diesendorf, 2007, p. 251). Insurance against future meltdowns (by accident or terrorist takeover) or bombings, involving hundreds of thousands of casualties, will be astronomical. When proponents talk of nuclear power being cost-effective, they simply ignore all of these costs, leaving them to governments and future generations to worry about.

Uranium enrichment and reprocessing can both produce explosives for nuclear weapons in the form of further enriched u-235 (possibly as low as 20% is sufficient) and 'weapons grade' plutonium. 'An ordinary 1000 mw nuclear power station produces about 200 kgs of reactor grade plutonium, enough for at least 20 nuclear weapons' (Diesendorf, 2007, p. 262). Over the years, many countries' 'civil' nuclear programs have also involved secret military programs for development of nuclear weapons. Further proliferation of civil nuclear power programs would encourage and make possible more such military programs, threatening wars which would kill hundreds of millions of people.

Planning and construction of a nuclear power station takes 10 years, using masses of fossil energy, as against safer, cheaper, sustainable alternatives that can be up and running much more quickly at much lower energy cost. Also 'even with a slow expansion of the global nuclear industry, there is likely to be a global shortage of high grade uranium over the next few decades' (Diesendorf, 2007, p. 268).

There are other technical possibilities including the accelerator driven thorium reactor which could, according to some experts, be safer than standard reactors and produce safer waste materials. There is also the possibility of fusion power – generating energy by fusing the nuclei of light elements such as hydrogen, deuterium and tritium to form heavier elements. In theory heavy hydrogen could be extracted from sea water and, combined with lithium, used to produce energy with no emissions, the only waste product being the non-radioactive gas helium-4.

Experiments have continued for 50 years to try to create a contained nuclear fusion reaction. As Maslin says some advances have been made at the Joint European Torus project in the UK which has produced 16 megawatts of fusion power. The problem is the amount of energy required to generate the huge temperatures in the first place and the difficulty of scaling it up to power plant size. (2009, p. 152)

An international project, ITER, is underway to construct a bigger torus – or tokamak – capable of producing 10 times as much energy as it consumes. This will be used to develop ways to effectively extract the energy – in the form of fast moving neutrons – rather than providing electrical power.

Compared to the large amounts of radioactive heavy metals produced by a fission nuclear reactor, or the tonnes of greenhouse gases produced by a coal burning power plant, fusion looks good. The problem is that proponents of fusion power tend to ignore the effects of the neutrons the fusion process produces. Neutron radiation makes the whole complex containment vessel, of electromagnets, exhaust pumps and blanket components, highly radioactive. Fusion will never be the cheap, clean, safe energy source it was once advertised to be.

Commercial fusion reactors are not possible until 2045. If fossil fuels are still supplying energy at that stage, huge amounts will be needed to construct, shield and decommission these complex and dangerous facilities.

The endless endorsements of clean coal and nuclear energy emanating from political and business leaders, whether taken seriously as genuine 'plans' for the future, or merely as diversions from funding, or even discussing, safe and effective possibilities, seem to offer solid support for Herve Kempf's (2009, p. 94) claims that such leaders 'harbor an unconscious desire for catastrophe'. This is all the more depressing given the fact that real, safe and relatively cheap solutions are available, and time is rapidly running out.

DISCUSSION TOPICS

1. On the basis of the evidence presented in this chapter, are current policies capable of averting catastrophic climate change?
2. What could be done to avert such change? What are the obstacles to effective policies in this area and how could such obstacles be overcome?
3. Is nuclear power the answer?

Health care provision

Private provision

Previous chapters have considered neoliberal endorsement of private production and free market distribution of the necessities of life: food, water and housing, along with energy generation and transmission, which provide the foundation for good health. Such private provision is deemed both more efficient and more just than any public alternative.

This chapter looks at private provision of basic medical services. Here again, neoliberals support user-pays for services from private health providers including hospitals, medical centres, surgeons and GPs, directly to such providers, or through voluntary private insurance provision. Medical providers should be free to offer their services at prices they think are reasonable to whoever is prepared to pay for them. Medical consumers should be free to choose to buy such services directly or to buy insurance from private providers. Private corporations should be free to offer either or both sorts of services if they choose to do so. If health care is seen as a commodity to be bought and sold in a competitive market, with supply and demand determining how it is organised, financed and delivered, the rational actions of producers and consumers will automatically create a system that delivers health care as efficiently as possible.

From the neoliberal perspective, public funding for health care is bought at the expense of the state taking money (as tax) which could be used to drive efficient market processes. Such funding is channelled into intrinsically inefficient systems of public provision. Consumers have to accept the services offered by a single state monopoly rather than choosing a provider best able to serve their needs from a range of possible competing alternatives.

Rather than being able to choose to spend what they want on their own health care, consumers are forced to spend potentially large amounts on other people's health. The progressive taxation systems favoured by social democrats to finance public provision force higher earners to pay the health costs of those who contribute much less – or nothing at all – to national wealth. Money, which should have served to motivate further efficient production, is taken away from those who earned it and given away to those who did not. So non-taxpayers will inevitably abuse the system. A market based system does a much better job of controlling costs and maintaining the quality of care.

Milton Friedman's extreme laissez-faire position rejects any kind of regulation or intervention by the government or professional bodies. Even the most basic licensing of medical personnel by central authority – to guarantee to consumers that such personnel have medical training and competence – is rejected as government mandated monopoly power which reduces supply and increases prices. Market forces alone should be left to drive out incompetent doctors as consumers will soon discover which ones are the good practitioners.

Other neoclassical theorists have identified problems in equating medical services with other commodities as objects of operation of free and efficient market forces. In particular, individuals don't necessarily have knowledge of the competence of medical personnel, particularly specialists upon whose skills their life might depend. They also do not know the real costs of medical procedures.

Kenneth Arrow has argued for the necessity of entry barriers to reduce such uncertainty. Licensing requirements guarantee to consumers that doctors have appropriate training and competence. Rather than direct government intervention, Arrow looked to physicians' self-regulation based on professional ethics and norms as the best correction for 'informational asymmetries' in the market for health care. Doctors themselves are in the best position to formulate and enforce basic standards of medical competence. In fact, only doctors themselves really understand such things (Gostin et al., 2005, pp. 631–2; pp. 645–6).

Arrow also highlighted issues of 'moral hazard' in relation to medical insurance. People with insurance are more likely to take serious health risks or demand more services than they otherwise would in the knowledge that their medical expenses will be largely covered by others (Pressman, 1999, p. 180). This will lead to increasing demand for health services with rising insurance premiums and health care spending. This problem is, of course, even worse with tax funded – apparently free – state medical services. The solution is co-insurance, with individuals paying a large proportion of their own health care costs.

Another issue is that that of 'adverse selection' – where those with severe health problems have a strong incentive to keep them secret from insurance companies and buy lots of insurance. As Pressman says

> If insurance companies set rates based on average risks, high risk groups will purchase a lot of insurance and low risk groups will buy little or no insurance. The companies will lose money and raise rates. This will drive out more low risk groups. (1999, p. 181)

Arrow's solution is for everyone to be covered by health insurance so insurers don't have to worry about low risk individuals dropping out of the system. Another option, favoured by insurers, is to force insurance purchasers to disclose all relevant risks (pre-existing conditions, genetic propensities etc.) to them when buying insurance and to provide them with legal rights of non-payment with evidence of non-disclosure. It is fair, just and efficient that insurers should have the right to exclude high-risk individuals, or charge them higher premiums, in proportion to the extent of the risk involved.

Generally, proponents of laissez-faire argue that free markets will maximise consumer choice, of high cost, high service policies or low cost, low service policies designed for individual needs, with providers and insurers forced to compete on price and quality to retain customers and price their services in proportion to real costs. Patients can then make the same free and rational decisions about their health care expenditure as they do when shopping for cars or DVD players.

There are difficult questions about the possible provision of a minimum government safety net of publicly funded basic services for the very poor who cannot or will not pay for services. This is justified by the requirement to avoid the deleterious social consequences of lack of private medical cover, in the form of crime, social disruption and spread of disease. All neoliberals agree that such benefits have to be strictly means-tested to ensure that only those in desperate need are covered, in order to avoid 'perverse incentives' and significant economic inefficiency.

Neoliberal ideas in this area are strongly influenced by the neoclassical theory of public finance. The key idea here is that of market failure, with under-supply of goods that have positive externalities and oversupply of goods with negative externalities. The social benefits should be compared with the private ones in order to calculate the amount of subsidy or charge required to bring the total quantity demanded and supplied equal to the social optimum.

An example would be a form of moderately effective immunisation. Some people will buy it to reduce the risk of infection, others will not, perhaps because of budget constraints – they would rather buy food. A particular pattern of supply and demand will determine how much vaccine is produced and what price it is sold for. The more non-buyers, the greater the possibility of major contagion, also affecting those immunised against the disease. To reduce the risk the price of the vaccine would have to be lowered, through state subsidies, to increase consumption. Alternatively, the state could allow and promote more complete markets, with those wanting to avoid contagion paying not only for their own vaccines but also for those who choose not to buy the vaccine for themselves (Mehrotra and Delamonica, 2007, pp. 97–8).

We can see how these ideas apply to basic medical services in general, as well as vaccinations in particular. It makes utilitarian sense for some minimum subsidies to address infectious diseases and other socially disruptive conditions of the poor as well as for mandatory public health measures of disease control, including enforced confinement and treatment of some conditions.

Free choice

Chapter 3 began to develop a comprehensive critique of neoliberal economic ideas. At this stage, we merely raise some obvious issues in relation to the neoliberal market based approach to health care. In particular, the fact that people do not choose to get sick, in the way in which they might choose to get a bigger car, and therefore a market in health care is rather different from a market in cars or other consumer goods.

It's true that people don't choose to get hungry or cold either, but they need to buy food and shelter, and in the absence of adequate public transport, they frequently need to buy a car to get to work to earn the money for food and shelter. This is an argument for separating the provision of food, shelter, health care and public transport, from more genuinely 'free' consumption choices and utility calculations. It is a dangerous mistake to see food, shelter or health care as options or choices to be balanced against other consumption choices. Rather these are structural preconditions of the possibility of being able to make meaningful choices. As such, they need to be treated as universal human rights, which cannot be traded off against any other choices or against protection of private property.

There is nothing about market relations that guarantees adequate resources for the satisfaction of such basic needs of all will be provided at prices that all can afford. There are no guarantees of the necessary expectations of profit required to drive such provision. Universal and well paid employment to enable everyone access to such services cannot be guaranteed. On the contrary, all the indications are that unassisted market relations are quite incapable of delivering any such outcomes. Such profit driven market relations have never managed to provide high-quality health care for more than a wealthy minority.

In this context, it makes sense to ensure communal, non-market provision of goods needed by everyone for any kind of worthwhile human life, through a fair labour contribution by everyone, while allowing individuals to earn extra money so they can choose to acquire goods that they merely want. Here then, we distinguish a world of individual choice, with greater social labour contribution rewarded by greater scope of choice, from a world of planning in the service of need.

Neoliberals frequently counter-pose the 'free choice' made possible through market relations, with the 'rationing' implied by public provision and black markets. They play upon the fact that, for many, 'rationing' conjures up pictures of wartime privation, or of older person X deprived of life-saving access to medical equipment so that younger person Y can use the equipment in question.

At the same time, they argue that, without free market relations, which offer people the opportunity to work to pay for the standard of health care they want, such universal care can be sustained only through conscription of the necessary labour, or through intrinsically unfair and excessive taxation, which will, in turn, threaten overall wealth creation.

Public provision just means rational planning, including both effective job creation and a fair taxation system, with scope for equal access to high-quality services for everyone.

The market rations goods on the basis of ability to pay rather than need or merit. In relation to health care, this kind of rationing means avoidable death and ill health for those who cannot pay. 'Rationing' of public health care, by contrast, if done well, can mean that high-quality medical services are available to all if and when they need them.

We are frequently told that people do, in fact, choose their health status by virtue of particular 'lifestyle' choices such as eating, smoking, failing to use the safety equipment provided etc. This issue has been discussed in Chapter 4. We have seen how the major determinants of individual health are, for the most part, beyond the reach of individual choice and action; certainly of the choice and action of the poorer majority. Individuals have very limited scope to 'choose' their social and economic circumstances, the level of stress to which they are subject at work and at home, their early life experiences, degree of social exclusion, poverty and relative deprivation, their level of job security or addiction to alcohol, drugs and tobacco.

In some cases, those who argue strongly for 'individual lifestyle choices' as causes of ill health are the same people who profit from selling addictive, but legal, drugs; who pay advertisers to encourage children to consume junk foods; who produce and market dangerously fast cars; who fail to provide safe working environments for those they employ; and support legislation that causes workers to be fined or sacked for stopping work because of inadequate safety provision. There are solid grounds for arguing that it is almost always radically misleading to try to understand health outcomes in terms of such individual lifestyle choices. At this stage, it's only necessary to recognise that health outcomes are not always or entirely a product of such 'free choice'; that in a significant number of cases, individuals suffer from some illness or accident through no fault of their own. It is ethically unacceptable to punish such individuals with (avoidable) disease and death by virtue of their inability to pay for private services.

Insurance issues

The advent of the possibility for an increasing range of genetic tests has thrown into sharp relief many of the issues and arguments of the free marketeers.

Insurance companies campaign for the right to demand their own comprehensive testing of all applicants, the right to exclude individuals at particularly high-risk, or charge them much higher premiums. They sometimes appeal to a principle of 'actuarial justice' to the effect that 'people who have similar health or similar life expectations should pay similar premiums – while those who have poorer health or lower life expectations should pay more (Rouvroy, 2008, p. 205). This principle is itself 'justified' by the argument that it is unfair that those at low risk should pay for those at high-risk. If governments enact laws of genetic privacy or genetic non-discrimination, as in Article 11 of the *Council of Europe's Convention on Human Rights and Biomedicine*, then all premiums will have to be so high as to make insurance unattainable or unaffordable for many. 'It is not just insurance companies who stand to lose if a person with a debilitating gene is included in the insurance pool, so, too, do the other insured' say Buchanan et al. (Rouvroy, 2008, pp. 206–7)

Of course, in terms of any genuinely ethical idea of justice, any such 'actuarial' justice is totally indefensible. As Rouvroy (2008, p. 206) points out, there is no way that those already disadvantaged through no fault of their own, should then be further disadvantaged by being denied health insurance cover or forced to pay exorbitant premiums.

There is an attempt here to defend such actuarial justice in utilitarian terms. Without full disclosure and discrimination the majority will suffer through higher costs or lack of available private cover. In particular, the argument is supported by reference to 'adverse selection' as considered earlier, with those who know they are at high-risk taking out high levels of health and life insurance cover. As Rouvroy points out, such arguments are not supported by available data. She says women who tested positive for the BRCA1 gene mutation did not capitalise on their information advantage by purchasing more life insurance than . . . women who have not undergone genetic testing. (Rouvroy, 2008, pp. 209–10)

It is perfectly possible to address the issue through a 'threshold' system of insurance cover below which generic information need not be disclosed – as in the UK and the Netherlands (Rouvroy, 2008, pp. 199–200) – providing, of course, that such cover is really adequate.

As Arrow acknowledges, the threat of being deprived of insurance at an affordable price motivates individuals to avoid medical tests. These tests could be valuable in preventing or mitigating the adverse consequences that they might predict. Without them, there could be worse health outcomes and potentially greater pay-outs by insurance companies.

At the same time, the whole idea of genetic testing as a basis for discrimination is misguided. Only a tiny percentage of diseases are caused by the mutation of single genes (Rouvroy, 2008, p. 210). Most are caused by the interaction between many genes and other bodily and environmental factors. The illnesses actually caused by failure of operation of a single gene will themselves probably be ameliorated by effective treatments in the future, which need not necessarily be exorbitantly expensive (though this will depend upon other areas of health care policy).

As Rouvroy (2008, p. 196) says, 'such genetic individualisation is antithetical to the basic function of insurance which is precisely the collectivisation of the burden of individuals' risks'. The logical conclusion of the insurance companies' position is that they make substantial profits by insuring only those who will never make any claims, while ill people have to pay the full costs of 'free market' treatment.

On the one hand we are told that insurance companies' profits must be preserved, since the erosion of public systems leaves only such private health care provision. On the other, we are told that their profits can only be preserved by excluding all people, at genuine risk of sickness, from private insurance (Rouvroy, 2008, p. 204).

This again highlights the vital need to get away from immoral and divisive private systems and move quickly to comprehensive publicly funded health care provision, giving each person what they need, built upon a foundation of fair contributions by all.

Health care in the US

In the developed world, the US comes closest to the neoliberal ideal, in contrast to other developed societies where the public sector is the main source of health funding. Amongst the Organisation for Economic

Co-operation and Development (OECD) countries only in the US, Mexico and South Korea is private health care dominant over public health care. In the US around 44% of health spending is funded by government revenues, well below the OECD average of 72%. In 2003, 62% of non-elderly Americans received employer sponsored insurance, and 5% purchased insurance on the private non-group (individual) market. Fifteen per cent were enrolled in public insurance programs like Medicaid, run by the states, and jointly funded by state and federal government for low income and disabled people. A further 18% were uninsured, potentially purchasing medical services direct from the providers (National Coalition of Healthcare: http://nchc.org/facts-resources/quick-facts/18-americans-under-age-65-have-no-health-insurance).

Individuals aged 65 or over are mostly enrolled in the federal Medicare program, funded by a special payroll tax. While around 34 million senior citizens and five million disabled people are entitled to hospital services, they have to pay a monthly premium of $54 to be covered for physician services and preventive care. Many purchase additional private insurance to cover health care expenses not covered by Medicare (Krugman and Wells, 2006).

Unlike the majority of working Americans with private health insurance, Medicare beneficiaries are free to seek medical care wherever they choose. Partly as result of this, and the increasing cost of drugs, the system faces serious funding shortfalls, and George Bush moved the system closer to privatisation in 2003, reducing coverage for out of pocket expenses and increasing premiums while at the same time preventing the federal government from using its power to reduce drug costs.

Employer sponsored insurance represents the main way in which Americans access health insurance. Employers provide health insurance as part of the benefits package for employees. They deduct health insurance premiums from their tax as a cost of doing business but not all employees are covered. For example, only about half of Wal-Mart's 1.4 million employees are covered by health care benefits (http://walmartwatch.com/issues/health_care).

Insurance plans are administered by private companies, both for profit, for example, Aetna, Cigna, and not-for-profit, for example, the BlueCross BlueShield Association or else the company pays the health care costs of employees directly. While employers usually pay the majority of the premium, employees typically also contribute (now an average of around US$3300 per year). The benefits vary widely with each specific health insurance plan. Some plans cover prescription drugs while others do not. Some cover preventive care such as children's vaccinations, flu shots and eye examinations, other (including Wal-Mart) do not.

Again, this accords with neoliberal ideas of choice, incentives and competition. Employers can choose the fund they use, and this can influence an employee's choice of employer. Employers can attract the quality of workers they need, in part through their choice of fund to offer them. Competition between private funds ensures high quality of service at minimum cost.

History

Prior to the 1965, doctors controlled most aspects of the US health care system and had responsibility for quality of care. 'Healthcare was largely a local, private sector enterprise with minimal government involvement.' From 1965 to the mid 1980s, the federal government became increasingly involved as purchaser, provider and regulator. Doctors retained considerable autonomy and power as private practitioners. Most private health insurance coverage was provided through a fee-for-service model that allowed patients to visit the doctor or hospital of their choice.

Physicians were mostly solo practitioners with hospital admitting privileges (i.e. independent contractors) while insurers paid the bills. In fee for service medicine, based largely on employers' health insurance benefits, medical care and insurance were separate industries. As Gostin el al. explain

> Physicians, hospitals and other providers delivered health care. Separate entities, commercial insurers, [and] public programs such as Medicare and Medicaid . . . financed the care provided. Under commercial insurance plans, known as indemnity insurance, patients were charged for each service they received from the . . . provider. Patients either paid for their care out-of-pocket or their . . . insurance company paid for it. The insurer reimbursed whatever treatment the physician ordered based on customary fees. (2005, pp. 623–4)

As Gostin et al. point out, physicians recommended tests and treatment whenever there was an expected benefit, regardless of the costs. Patients were not exposed to the true costs of care and as a consequence, say neoliberals, health care costs continued to rise rapidly as medical technology developed. From the mid 1980s employers began, increasingly, to hire 'health plans' to 'manage' their employees' health care by controlling access to care and lowering costs. The federal government encouraged the development of such 'managed care' as a market based means of cutting costs, through combining the financial and clinical aspects of medical care in a single entity – the managed care organisation. By the beginning of the 1990s managed care had replaced fee-for-service as the dominant means of health care provision.

> The managed care revolution . . . combine[s] the financial and clinical aspects of medical care into a single entity . . . [It] compels physicians to balance between conserving assets for the MCOs patient population and providing care to their individual patients . . . managed care was exciting to many . . . because it promised to stem spiralling . . . costs by placing constraints on the

amount of care provided, without sacrificing... quality... combining the medical and insurance functions into one entity provided strong incentives to reduce... costs. It also introduced price competition into the markets for health insurance and health care services. (Gostin et al., 2005, p. 624)

Gostin et al. point out that managed health care has subsequently evolved in its organisation and in the 'extent of its containment initiatives'. In some cases, physicians are employed directly by the Health Maintenance Organization (HMO) and deal exclusively with HMO members.

Providers are 'striving to build a tightly coupled organisation that embraces all aspects of patient care, including financial integration, physician integration and clinical integration'. The goal is to link multiple provider organisations... to have one organisation offer the full range of patient services. The more tightly integrated... the easier it is to market the system to employers and insurers. Through horizontal integration corporate systems are able to expand the scale of their operations. Through vertical integration they are able to increase the scope of the services provided. (2005, p. 626)

Managed care puts administrators and designated 'gatekeepers' (usually primary care physicians) in charge of 'guiding' patients through a health care network, with the goal of managing costs, through the most efficient use of available resources. Patients are required to check with the health plan for approval before visiting a specialist or receiving a medical procedure. In practice it means restricted access to doctors and procedures.

The system encourages the use of less highly trained providers for less demanding services and provides financial incentives for physicians to limit their utilisation of care, for example, year-end bonuses or holdbacks of payments received only if they do not exceed specified utilisation limits. In addition, managed care increasingly employs data from outcome (efficiency) studies to develop practice guidelines and for the ongoing assessment and refinement of diagnostic services and treatment services.

Some neoliberals justify the managed care system in terms of practicality and efficiency. By reducing the rate at which the cost of health care is rising, managed care keeps insurance more affordable than it would otherwise have become. Others criticise managed care on the grounds of managed care organisations restricting the fee choices of consumers, 'skimming the cream of patient populations', taking on only relatively low cost, healthy patients and pressuring physicians to ration care, thereby interfering with their professional fiduciary obligation to provide the best care for each patient.

Defenders of the US system highlight the fact that adequately insured individuals, in 'appropriate locations', can still 'receive the best healthcare in the world', with good access to tertiary care hospitals with the most modern equipment for both diagnosis and treatment. Modern drug therapy, telemetry, and highly skilled professionals are readily available. The US is a world centre for the most sophisticated surgical services, and most procedures can be performed with minimal delay. In one or two areas, such as breast and cervical cancer survival, the US is clearly the world leader. (Uretsky, 2009)

Other OECD systems

During the period of social liberal consensus, particularly from the 1950s to the 1970s, the proportion of public expenditure devoted to welfare benefits and services rose throughout the OECD, along with state regulation of labour markets. By the end of the 1970s, public expenditure on welfare, devoted principally to income maintenance programs, health care services, education and housing, represented 25% of gross domestic product (GDP) in the UK, and 33% of GDP in Denmark, Sweden and the Netherlands.

As Mappes and DeGrazia point out, Canada funds almost all health care through the federal and provincial governments.

> Providers are either paid a fixed sum for each patient enrolled in their practices or they submit ... standardised forms and get paid on a fee for service basis. While some private insurance companies offer elective services not covered by the universal Medicare plan [e.g. cosmetic surgery] the role of private insurance is marginal, due in part to the illegality of private companies providing the same services that are offered in the universal Medicare package. (2006, p. 624)

The Canadian system provides quality health care to all and achieves indices comparable to those of the US while spending only slightly more than half as much per capita – US$2250 versus US$4270 in 1998. Such provision includes 'a comprehensive package of healthcare services', including 'in-patient and ambulatory care, long-term care for chronically disabled elderly, and psychiatric services'. Some provinces also cover dental and chiropractic care, optometry and prescription drugs.

> The administrative simplicity of the Canadian system allows it to spend only one eighth of every health care dollar on administration while the US spends one quarter of each dollar for this purpose. (Mappes and DeGrazia, 2006, p. 624)

A majority of Canadians are strongly supportive of the system. However, decreased public health expenditure in the 1990s led to longer waiting lists for elective surgery. As a result, some wealthy Canadians cross the border to get immediate access to high tech services in the US, rather than waiting for them in Canada. At the same time, a number of US states purchase drugs in

Canada, where effective price controls make many drugs significantly cheaper than in the US (Mappes and DeGrazia, 2006, p. 624).

In Germany, not-for-profit, semi-private 'sickness funds' offer coverage to all citizens.

> Sickness funds set premiums based on one's ability to pay and not on risk factors such as occupational risk or one's health history. Members of sickness funds choose their own physicians. The . . . coverage includes inpatient and ambulatory care, prescription drugs, dental services, psychiatric care and chiropractic and optometric care; coverage also includes cash benefits . . . [for maternity, necessary travel, burials]. At the same time long term care . . . is means-tested and some . . . services require modest co-payments by patients . . . Hospital doctors and nurses are salaried. (Mappes and DeGrazia, p. 625)

One tenth of the population pays for for-profit private insurance or is provided with government insurance. Surveys suggest that most people are reasonably happy with the quality of care and freedom of choice within the not-for-profit system, with equal access for all citizens at a cost far below that of the US system.

Founded in 1948, the British National Health Service (NHS) provides universal medical, dental and nursing care for all, funded by the state through tax contributions. Private care and insurance were not abolished.

> British citizens may choose to pay for private insurance, while their tax contributions fund the NHS. Unlike their Canadian counterparts, British physicians may provide a particular type of medical service both within the public system and outside it, and private insurance companies may cover services that are provided in the NHS. (Mappes and Degrazia, 2006, pp. 625–6)

The NHS offers in-patient, ambulatory and long-term care, along with dental services, prescription drugs (with a co-payment), psychiatric care, chiropractic care and optometric services. NHS doctors are salaried or receive a fixed sum for caring for individual patients. However, patients have a limited choice of primary care physicians, whose referrals they need to access specialists. Aggressive cost limitation measures, leading to long waits for routine surgery, outdated equipment and simplified administration kept UK health spending per capita well below that of Canada for some time ($1813 versus $2580 in 2000). Spending on the public system has subsequently increased and public support remains strong (Mappes and DeGrazia, 2006, p. 626).

Neoliberal critique

Many neoliberal discussions focus on what are seen as the major weaknesses of the beleaguered and disintegrating public services that still exist around the developed world as survivals from a previous age of social liberal reform. Increasing government debt, burgeoning costs of new treatments, over-servicing and an ageing population are all seen as insuperable obstacles to the maintenance of any sort of comprehensive public funding of health care.

Considering the Canadian situation, for example, Dirnfeld highlights the 'excesses of planners and dreamers' which have allowed Canadian public health care, 'like other social programmes' to grow 'too comprehensive and too costly to be funded by the public purse'. 'The effects of – considerable – cutbacks in health care funding' have included

> closure of hospitals . . . elimination of some programs and significant curtailment of others . . . and stagnation of programs that needed to grow to keep pace with a growing and aging population . . . or with improved technology . . . Canadian Medicare has resulted in the tyranny of a monopsony, which has led to rationing through the use of queues, to decreasing accessibility and to diminished quality. (1996, p. 408)

In this context, neoliberals argue for the need to develop or expand a 'complementary' private system. As Dirnfeld says

> the private option – of insurance with premiums adjusted for actuarial risk – serves to decrease waiting lists for public facilities, thereby improving access for those using the public system. The private system offers a standard of care to which the public system must aspire . . . Without a parallel, private system, the public system delivers care according to what the government payer dictates – and that can mean . . . rationing and decreasing quality of care. The cost-effectiveness of the private system has been demonstrated . . . Experience worldwide has shown that 10% to 15% of the public opts for a private system. Therefore there has been no significant movement of the best physicians and nurses from the public system to the private one, because there is not enough work in the private system alone to support them . . . Choice is a cherished value. Denial of choice in medical care as in other areas of human experience, is unacceptable to many. (Dirnfield, 1996, p. 409)

The developing world

As Mehrotra and Delamonica observe, some developing countries provided universally available health services for all, paid out of government revenue, at the time they achieved independence from colonial rule. Richer citizens still tended to buy private health insurance or buy medical services directly, in the absence of such insurance, as in Sri Lanka and Kerala (Mehrotra and Delamonica, 2007, p. 248). The World Bank has, for decades, encouraged further expansion of the private system. Structural adjustment programs in the 1980s required cost recovery and user charges in much of Africa and Latin

America, with user fees decreasing utilisation of medical services (Mehrotra and Delamonica, 2007, p. 249).

> Since the mid 1980s many countries that had limited or that had entirely banned for-profit practice – e.g. Malawi, Tanzania and Mozambique – have been encouraging private providers by regulatory liberalisation and fiscal incentives – [in some cases including] privatisation of public hospitals. (Mehrotra and Delamonica, 2007, pp. 247–8)

As Mehrotra and Delamonica point out, throughout the developing world, there are four main organisational forms of health service provision.

> One is the ministry of health, with a network of public providers, financed from general taxation. A second is private health insurance based on payment of premiums. The third consists of social security organisations based on salary related contributions, usually for formal sector workers. Finally, there are community or provider based pooling organizations. (Mehrotra and Delamonica, 2007, p. 171)

The second and third forms are only of significance in economies with a large formal sector. In most low income countries, for the poor, who are usually rural farm workers, the realistic options are public finance and public providers, or some form of community pooling of household resources to meet the costs of public/private provision or out-of-pocket payment to private providers.

In most cases, users of health services make direct payments at the time of accessing services, 'usually formal payments to private providers and often informal payments to public providers' (Mehrotra and Delamonica, 2007, pp. 168-9). Out of a total of around US$85 billion spent on health care each year around the developing world, 'half comes from patients' pockets and in the case of the poorest 32 countries in the world nearly 60% or US$6 per capita' (Mehrotra and Delamonica, 2007, p. 169).

Mehrotra and Delamonica (2007, p. 169) give figures of around two thirds private – household – contributions for Asia, including 75% in India. In sub-Saharan Africa the figure is 39%, with aid donors supplying 33% and governments only 28%. They argue that 'economic reforms' in China, with big cuts in public expenditure, have effectively destroyed public health care, and substituted a user-pays system. Public hospitals now operate on a 'fee for service basis' (Mehrotra and Delamonica, 2007, p. 250). Members of farming families are forced to seek employment in the cities in order to earn enough to pay other family members' medical costs.

The Americas is the only region where public spending on health exceeds that of direct household payments. Latin America experienced a massive transnationalisation of its health sector in the 1990s with US multinational managed care corporations Aetna, Cigna, AIG and Prudential, entering insurance and health services, aiming to 'assume administrative responsibilities for state institutions and to secure access to medical social security funds' (Mehrotra and Delamonica, 2007, pp. 248–9).

These corporations buy or enter into joint ventures with already established indemnity insurance, or prepaid health plan providers, or agree to manage social security and public sector institutions. Encouraged by the World Bank and the International Monetary Fund (IMF), 'penetration by multinational corporations in health of these social security funds is most advanced in Argentina and Chile' and is growing in Brazil and Ecuador (Mehrotra and Delamonica, 2007, p. 249).

The World Bank's *World Development Report 1993* argued that

> the primary objective of public policy should be to promote competition among providers – including the public and private sectors . . . competition should increase consumer choice and satisfaction and drive down costs by increasing efficiency. Government supply in a competitive setting may improve quality or control costs, but non-competitive public provision of health services is likely to be inefficient or of low quality. (Mehrotra and Delamonica, 2007, p. 249)

Here again, neoliberals highlight the efficiency and quality of private sector provision compared to the public sector. The way in which an expanding private sector reduces pressure on an overextended public sector, thereby freeing up capacity and resources in the system as a whole, is applauded. The International Finance Corporation of the World Bank, for example, defended its support for the expansion of the private sector on precisely this basis.

Assessment of private health care in the developed world

Rather than providing the ultimate vindication of neoliberal policy in the area of health care provision, the US system is actually a massive indictment of such policy. The US is the wealthiest nation of the OECD (the organisation of industrialised countries) and spends more on health care per capita than any other OECD country. In 2003 total health spending per capita was US$5635 (adjusted for purchasing power parity), more than twice the OECD average of US$2307. Between 1998 and 2003 health spending per capita in the US increased in real terms by 4.6% per year on average, a growth rate comparable to the OECD average of 4.5% per year. However, in 2002 the US had only 2.3 practising physicians per 1000 population compared to the OECD average of 2.9; only 7.9 nurses compared to 8.2 in the rest of the OECD, and 2.8 acute hospital care beds, compared to the OECD average of 4.1 (http://www.oecd.org/dataoecd/15/23/34970246.pdf).

While average life expectancy in the US increased by 7.3 years between 1960 and 2002, it increased by 8.4 years in Canada over the same period and 14 years in Japan. In 2002–03, life expectancy in the US stood at 77.2 years, well below the OECD average of 77.8 years. Japan, Iceland, Spain, Switzerland and Australia all had higher life expectancy. In 2002, infant mortality in the US was 7 deaths per 100 live births compared to the OECD average of 6.1, and below 3.5 in Japan, Iceland, Finland and Norway. Adult obesity rates were the highest in the US in 2002 – at 30.6% compared to 24.2% in Mexico (in 2000) and 23% in the UK (in 2003.) (http://www.oecd.org/dataoecd/15/23/34970246.pdf)

The US system is highly inefficient and wasteful with more than 1000 different organisations paying health care providers. Most of them, as for-profit private providers, spend large amounts of money on advertising, determining patient eligibility, enforcing complex restrictions on coverage, conducting patient by patient utilisation reviews, attempting to collect bad debts, and paying huge executive salaries.

The US system is highly inequitable. Those who receive health insurance at work receive a tax break worth more than $130 billion a year. Meanwhile, many low paid workers find it difficult to obtain affordable coverage because their employer does not include health insurance as a workplace benefit (Irvine, 2002). In practice, most consumers are unable actively to shop around – in a free market – for insurance plans. If they do not like the health insurance and restrictions offered by their employers, they face adverse tax consequences if they want to buy health insurance on their own outside the workplace. The quality of care patients receive can differ radically depending on the type of insurance they have and on where they happen to live. As Uretsky says

> Patients in rural areas where there is a shortage of all types of healthcare personnel and patients who have no or inadequate insurance coverage get care below the national standard. (Irvine, 2002)

As vividly portrayed in Michael Moore's powerful film *Sicko*, a doctor who has acted as gatekeeper for managed funds testifies to the US Congress about the dozens of patients she has condemned to death by failing to allow them access to expensive treatments. Such access is denied through claiming that such treatments are 'still experimental', unproven, ineffective, or inappropriate for the given case (Moore, 2007).

In 2000 an estimated 38 million Americans did not have health insurance. With health insurance so closely tied to the workplace, moves from one job to another typically leave individuals and families without health cover. Buying health cover in the individual market can be very expensive in many states and individuals decide instead to buy food, housing, transportation and other necessities.

It's true that any hospital in the US that accepts Medicare or Medicaid is legally bound to provide medical treatment and stabilise any patient who presents with a medical problem, whether or not that patient can pay the bill. Hospitals that treat a substantial number of poor people receive a disproportionate share payment from the federal government but the uninsured get little or no follow up care beyond the immediate hospital emergency room. They often wait until later stages of illnesses to get medical care. And they also live in fear that they or their family members will get sick or have an accident and that the bills would leave the family in debt bondage forever.

Analysis

Neoliberal governments have run down public provision to the point where it is more dangerous, or perceived as more dangerous, for some ill people to enter public hospitals than soldier on outside. Endless waiting lists mean that seriously ill and disabled people suffer for years, becoming progressively worse, or even die, before being allowed to receive hospital treatment. This has meant that more people have been forced to turn to expensive private services even if this has involved serious financial hardship and health-threatening extra work, anxiety and stress. Medical personnel inevitably abandon a disintegrating public system in favour of better working conditions in an expanding private sector, causing even more problems for the public sector.

Such governments have claimed that more of the public 'prefer' the free choice of the private system. Along with the now standard references to ageing populations, overservicing and exponential cost increases, such a claim is used to justify further running down of the public system. Neoliberal politicians emphasise that, whatever the level of public provision, there are no limits to 'over use' by individuals with infinite wants for medical services, so long as such individuals don't have to pay in proportion to the real costs of their own treatment.

This is not only misleading in itself, as indicated by earlier considerations, but it also neglects some centrally important issues. One such issue, in relation to keeping down cost increases in a public system, is the possibility of keeping people healthy and happy so they don't need to access expensive curative health services – prevention rather than curative intervention. As argued in Chapter 4, with provision of basic food, water, sanitation, education and habitation, social inequality, stress and disempowerment resulting from significant inequality become major factors in determining levels of morbidity and mortality. Increasingly alienated, lonely, fearful, confused and oppressed people are in need of more psychological support. Addressing such issues has the potential to significantly reduce health care costs. Fear of the costs of private health care is itself a significant source of disease producing stress, so there is important feedback between prevention and cure.

A significant part of the cost increases of health care systems around the developing world in recent decades has come from spending on drugs. This is partly a function of increased use of drugs, particularly newer drugs, and partly of the increasing cost of new drugs. These issues are addressed in detail in Chapter 9.

Many drugs – particularly new ones – are massively overpriced compared to their real production costs, including development costs to the producers. Regulatory failure has allowed the marketing of too many unnecessary, ineffective and dangerous drugs. Indeed, a substantial proportion of medical care now addresses the iatrogenic consequences of misuse of pharmaceutical products. Financial and psychological manipulation of medical personnel by drug companies has contributed to massive over- and mis-prescribing of such drugs. All of these things could and should be relatively easily and effectively addressed to keep public health costs down.

However, as considered in Chapter 9, neoliberal political leaders have moved in the opposite direction, increasing drug prices by extending drug patents, renouncing government powers to bargain for cheaper prices, and entering into 'free trade agreements' that undermine such government bargaining powers.

There is a broader issue of the 'false economy' (and the cruelty) of running down public provision in such a way as to deliver increasingly worse treatment outcomes across the board. This then requires further interventions to try to address them. Good quality care the first time round reduces the need for potentially costly follow up.

Neoliberals maintain that 'human nature' implies infinite wants for medical services as for other consumer goods. As noted earlier, alienated and frustrated people will indeed turn to medical personnel for help and for drugs. But a passion on the part of patients for excess consumption of medical services is an issue of public ethics and psychopathology rather than of human nature. There are few fulfilled people who really enjoy unnecessary visits to the doctor or unnecessary consumption of potentially hazardous drugs or endless unnecessary operations per se.

Those unable or unwilling to stop themselves from compulsive medication require psychological assistance. We must not confuse overservicing with rational use of the system. Second opinions and adequate investigations are crucial where serious health issues are involved, and any properly organised system allows for them.

Empirical evidence from countries like Sweden and France, with comprehensive public health provision, provides no evidence for the neoliberal claims. There is no evidence of overconsumption in a tax funded 'free' public service. The public does not prefer a 'free choice' if adequate public services are provided. As noted earlier, the public generally remains highly supportive of existing public systems even when they are allowed to run down.

As long as a private system is allowed to coexist with public provision in a nation or a region, particularly a region with high levels of economic inequality, it will pose an increasing threat to public provision, and hence to any adequate health care for the majority of the population. The rich and the powerful will have the incentive and the ability to shift limited health care resources from the public to the private system; such limited resources will follow monetarily effective demand. With a high-quality private system available, those rich people, who are also selfish, have no incentive to 'subsidise' public provision and they can conspire to reduce their own contribution to the public system or exploit its resources to subsidise their own private provision.

Making private provision illegal, by contrast, with the public system supported by genuinely progressive taxation, forces the rich and the powerful to ensure the highest possible standard of public care, provided in the most cost effective manner, because they will be using the public system and they will be paying for a significant part of it. They will not always have the opportunity to access a private facility somewhere else.

It is perhaps also worth noting at this point that similar arguments apply in relation to comprehensive public education as a basic pre-requisite for a healthy and a civilised society. More detailed consideration of the role of education in shaping patterns of health and illness has not been excluded from this book because it is not important. On the contrary, it is absolutely central to the issues considered here.

Private health care in the developing world

If comparatively rich Americans can't afford free market health care, it's easy to see what a complete disaster it is for poor developing world populations. Mehrotra and Delamonica (2007, p. 249) point to substantial evidence of 'market failure' in such private provision with extensive use of untrained staff and overservicing in the private sector, including unnecessary Caesareans in private maternity wards and costly overprescribing of drugs in poor areas. Privatisation encourages a shift from preventative to costly curative services, particularly drug use.

In many developing countries there is evidence of resources in the private sector concentrated on high tech equipment (for rich patients) with neglect of basic services.

> Where in Latin America managed care organisations have taken over the administration of public institutions, increased administrative costs [and returns to investors] have diverted funds from clinical services . . . [eg] administrative and promotional costs accounted for 19% of Chilean managed care [ISAPRES] annual expenditure [in 2001]. (Mehrotra and Delamonica, 2007, p. 251)

Rather than allowing the public sector to 'redirect' its scarce resources to those in need, there is evidence that the growth of the private sector drains personnel and other resources away from an already stretched public sector through its capacity to offer higher wages and better conditions. Most fundamentally, as Mehrotra and Delamonica point out

> the effect of privatisation . . . and the reliance on out-of-pocket financing is to worsen equity in health care. The most serious effect is that services are refused on account of inability to pay, and illness goes untreated. (2007, p. 252)

They explain how managed care organisations in Latin America have attracted healthier and wealthier patients with sicker and poorer ones shifted to the public sector and how the public sector employs stringent means-testing with 'rejection rates averaging 30 to 40% in some hospitals' (Mehrotra and Delamonica, 2007, p. 252).

Around the developing world, increasing numbers of people simply do not, or cannot, seek medical help when they get sick because of the cost (e.g. 35–40% of people in rural China). With average costs of hospital admissions equivalent to a substantial proportion of an average annual wage, families go increasingly into debt as a consequence of medical treatment (e.g. 20% of families in rural Vietnam).

The high cost of private medical treatment, along with heavy marketing of drugs from the developed world and the expansion of local drug production in some countries, has encouraged a 'huge growth in private pharmacies' in the developing world, with drugs now accounting for 30–50% of total health care expenditure, compared to 15% in the developed world.

As Mehrotra and Delamonica observe

> those who cannot afford professional services are essentially catered to by pharmacies, which often do not follow prescribing regulations. This is especially the case in South Asia, China and parts of Africa . . . Pharmacies have a financial incentive to overprescribe . . . which leads to unnecessary drug use and the development of resistance to drugs. (2007, p. 252)

Antibiotics, in particular, have been massively oversold and misused around the developing world (including their use in cases of diarrhoea), leading to widespread antibiotic resistance. In most developing countries there is little monitoring or effective regulation of the private sector – including such rampant drug abuse – by government health ministries. In some cases at least this reflects the much greater political power of the private providers themselves, than of victims, deprived of care altogether, overcharged and damaged by dangerous drugs.

India's situation is particularly dire, with minimal government regulation of medicine. Here a situation close to Milton Friedman's ideal unregulated free market actually exists, with some medical schools 'auctioning medical degrees', 'hiring fake teachers to fool inspectors that they had sufficient teaching staff' (Shah, 2006, p. 113), 'clinics operating out of residential flats, with kitchens turned into operating theatres' (Shah, 2006, p. 114). As Shah says

> neither the state medical councils nor the Indian Medical Association . . . police physicians' conduct. The Indian Medical Association espouses no code of ethics . . . and has fought off any imposition of minimum standards . . . regarding patient care or licensing. (2006, p. 114)

As Shah points out

> In 2004 there were over twenty thousand licensed drug companies in the country . . . flooding the market with over seventy thousand different brands of medicines . . . To regulate this . . . array of products the government employs just 600 drug inspectors . . . In a 2003 raid in Patna, in the eastern state of Bihar, 7 out of 9 drugstores were found to be operating without licenses. Pharmacists routinely sell prescription only preparations to patients . . . no less than one in four drugs available on the market is fake or substandard . . . . (2006, p. 115)

Public health care in Cuba

At the opposite end of the spectrum, medical care in Cuba shows what can be achieved through comprehensive state control of medical services, along with elimination of extreme poverty and radical inequality through economic planning in a poorer developing country.

Cuba has been denied the 'benefits' of neoliberal globalisation, deregulation and capitalist private property due to the US embargo and through the continuing commitment of its leadership to collective ownership and economic planning. As Cooper, Kennelly and Ordunez-Garcia (2006) point out, World Bank estimates in 2006 attributed to that country per capita income of less than US$1000 per year, with Cuba's own estimate, including various income subsidies or social wage components 'in the range of US$2000–5000 per year. Yet Cuba has achieved 'health status measures comparable to those of industrialised countries'. The generalisation of the Cuban model would 'transform human health in other poor and middle income countries' (Cooper, Kennelly and Ordunez-Garcia, 2006, p. 1).

As Cooper, Kennelly and Ordunez-Garcia (2006, pp. 6–7) show, Cuba has achieved effective control of infectious diseases, with elimination of 'polio, neonatal tetanus, diphtheria, measles, pertussis, rubella and mumps', control of dengue and containment of AIDS, with 'domestically produced triple therapy free to all since 2000'. There have been significant reductions in infant mortality compared to countries with a similar GDP, to a level '20% below the US [average] for all ethnic groups' in 2006. There has also been significant progress in control of chronic diseases, with 'a sustained downward trend in coronary heart disease' from 1982, 'with a slope close to the maximum achieved in Europe and North America (~1.5% per year)' (Cooper, Kennelly and Ordunez-Garcia, 2006, p. 1). Anti-hypertensive drugs and recombinant streptokinase are locally produced and widely used. Smoking rates fell by a third from 1986–2006.

While spending around 16% of its gross national product (GNP) on health – in 2006 – 'roughly $320 per year per person', Cuba has developed a comprehensive system of 'polyclinics and hospitals' and 'a family physician and nurse' on every block, providing care for 120–160 families' (Cooper, Kennelly and Ordunez-Garcia, 2006, p. 3). This system is supported by a developed public health surveillance system, including a 'high autopsy rate' and comprehensive statistical records as a basis for rational planning. Forty eight referral hospitals are equipped with magnetic resonance imaging (MRI) facilities and local polyclinics with ultrasound and endoscopy facilities. Dialysis and organ transplantation 'have been widely available for a number of years' (Cooper, Kennelly and Ordunez-Garcia, 2006, p. 6).

Cuba has managed to maintain significant levels of biotechnological research, producing the first meningitis B vaccine, the first synthetic antigen based flu vaccine and many other achievements (Cooper, Kennelly and Ordunez-Garcia, 2006, p. 4). Cuba's medical aid program had sent thousands of doctors, nurses, dentists and others to 52 countries by 2006. These people have committed to living and working in slums and disaster zones for two year periods. Cuba has trained tens of thousands more doctors – at little or no charge – for other poorer countries, particularly in Latin America.

Cooper, Kennelly and Ordunez-Garcia argue that effective central state control has allowed a 'strategic approach' to health care in Cuba, including immunisation of entire at-risk populations as soon as 'safe and effective' vaccines become available, with screening and treatment of susceptible populations in the absence of such vaccines. Local communities have been effectively mobilised to disrupt insect transmission of disease. Universal primary health care, with large numbers of trained professionals, has allowed a 'continuum of care' to developing infants and children, including

> pre-conceptional health of women, prenatal care, skilled birth attendants, and a comprehensive well-baby programme [reducing] infant mortality to levels approaching the biological minimum. (2006, p. 8)

As Cooper, Kennelly and Ordunez-Garcia (2006, p. 8) point out, 'these principles have been successfully implemented in Cuba at a cost well within the reach of most middle-income countries'. The very widespread failure of other such middle income countries and regions to implement any such programs, or to achieve anything close to Cuban levels of public health and public health care, is a major indictment of neoliberal globalisation and privatisation.

DISCUSSION TOPICS

1 How should health care be organised in advanced industrial societies?
2 Why haven't other middle income countries followed the Cuban example? Could they do so?
3 How should health care be organised in the poorest countries?

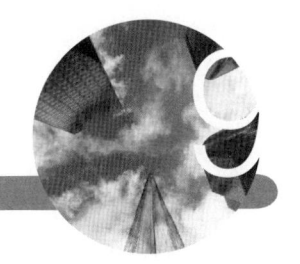

Drugs, drug testing and drug companies

The big drug companies today, and their predecessors, do not have an auspicious history. This chapter shows how recent developments have built upon and extended a long tradition of ruthless irresponsibility and criminality. It includes considerations of the subversion of the US Food and Drugs Administration (FDA), and the radical failure of that body to properly police the testing and sale of drugs.

Problem drugs

As Sonia Shah points out, in her book *The Body Hunters*, in 1885 the main ingredients in medicines sold by major drug companies were quinine and morphine

> Merck sold cocaine, Bayer sold heroin. Alcohol diluted with water was sold as a cure for colds, congestion and tuberculosis. Only after these unregulated medicines had killed thousands of Americans including countless infants who were given opiates, did Congress pass the 1906 Food and Drug Act, requiring drug-makers to list ingredients on their product labels. (2006, p. 37)

The first really significant breakthrough in drug treatment was the development of sulfanilamide in 1932, which prevented bacterial cells from multiplying, allowing the host's immune system to destroy them. It effectively tackled streptococcal infections, pneumonia, meningitis and gonorrhoea. However, by the late 1930s many strains of bacteria had developed resistance to it. Following the poisoning deaths of a hundred infants by a sulfa mixture, Congress passed the *Food, Drug and Cosmetic Act* of 1938, requiring toxicity tests of new drugs (Shah, 2006, p. 38).

Penicillin was introduced as the sulfa drugs lost their effectiveness, successfully targeting a range of bacteria, including those causing tuberculosis and syphilis. With the development of further potent antibiotics, public faith in the drug industry was hugely increased. As Shah observes, more public money was pumped into the National Institute of Health to provide new ideas and information for drug companies. By 1957 the drug industry had profit margins that were double the US national average – 19% of investment after taxes. New laws gave doctors control of more potent drugs with patients and insurers paying the bills (Shah, 2006, p. 39).

The fact that all was far from well in the drug industry at the time is clearly illustrated by consideration of some of the drugs widely prescribed in the 1950s and early 1960s. Here we consider just a handful of the vast number of problem drugs widely prescribed at the time, and some more recent developments.

DES

Steroid hormones, including both oestrogens and androgens, play a central role in the regulation of reproduction, and in the development of secondary sexual characteristics. Diethylstilbestrol (DES), the first 'orally active synthetic oestrogen', five times as potent as the strongest naturally occurring oestrogen, was produced by Charles Dodds and his collaborators in 1938 (Harremoes et al., 2002, pp. 90–1). As Ibarreta and Swan observe

> DES is inexpensive and simple to synthesise and the developing pharmaceutical industry quickly began worldwide production . . . under more than 200 brand names . . . following very limited toxicological investigation. (Harremoes et al., 2002, pp. 90–1)

It was touted as a treatment for a range of different conditions, including

> menopausal symptoms and prostate cancer, lactation suppression and as a morning after contraceptive pill. It was later used as a growth promoter in chicken, sheep and cattle . . . The use of DES for prevention of miscarriage was . . . promoted by the work of Olive and George Smith who conducted . . . uncontrolled trials of DES for use in pregnancy throughout the 1940s. (Harremoes et al., 2002, pp. 90–1)

It was then believed that decreasing oestrogen levels could produce spontaneous abortions, so increasing levels could prevent this. We now know that such decreased levels found in association with miscarriage were actually a consequence, rather than a cause. Common sense would have suggested the obvious dangers of prescribing a substance known to powerfully direct cellular differentiation to a pregnant woman. Thousands of doctors and patients seemed to lack any such common sense in this case.

Ibarreta and Swan note that 'in the early 50s two randomised placebo-controlled trials' led to reports of 'no statistically significant differences in adverse pregnancy outcomes when DES was compared to a placebo' (Harremoes et al., 2002, p. 93). However, doctors continued to prescribe the drug on a large scale, worldwide – even to women without previous pregnancy problems. An advertisement for one of the DES products read 'recommended for routine prophylaxis in all pregnancies' (Harremoes et al., 2002, p. 94).

A reanalysis of the 1950s' data published in 1978, 'after the long term carcinogenic and teratogenic effects of DES were known' (Harremoes et al., 2002, p. 94), found that these tests had actually shown DES to increase the risk of the end points it was sold to prevent. As the authors indicated, had the data been properly analysed in 1953, 20 years of unnecessary exposure to DES might have been avoided. The drug was widely prescribed for two decades after it had not only been shown to be ineffective but to actually produce the conditions it was supposed to prevent (Harremoes et al., 2002, p. 94).

'Reports that DES increased cancer incidence in laboratory animals appeared as early as 1938' (Harremoes et al., 2002, p. 91). Since then, further research has shown that it increases the incidence of cancer of the mammary glands, cervix and vagina in many different species

> In 1970 Herbst and colleagues reported . . . they had diagnosed a rare vaginal cancer in 7 young women, a cancer that had never before been seen in that age group in this hospital. The following year [they] published [evidence] that 7 out of 8 cases but none of 32 matched controls had been pre-natally exposed to DES. (Harremoes et al., 2002, p. 90)

Months later, the FDA 'withdrew approval for the use of DES in pregnant women' (Harremoes et al., 2002, p. 90). However, the use of DES by pregnant women continued even after 1971 outside the US. Total exposure 'worldwide has been estimated to be 10 million' (Harremoes et al., 2002, p. 92).

Follow up investigations found that 'DES daughters' suffered from a

> range of reproductive tract abnormalities . . . These included epithelial changes as well as structural changes such as narrowing of the cervical opening and uterine malformations. (Harremoes et al., 2002, p. 92)

These abnormalities were widespread amongst women whose mothers had been prescribed DES early in pregnancy, and 'resulted in serious reproductive consequences for those affected' (Harremoes et al., 2002, p. 92) including increased risk of ectopic pregnancy, spontaneous abortion and premature delivery. Sons of women who had DES also show increased rates of genital abnormalities. Studies in mice have shown increased susceptibility to tumour formation in the third generation, suggesting that DES grandchildren may also be at increased cancer risk.

As Ibarreta and Swan (2002, pp. 94–5) argue, 'if the 7 cases detected by Herbst had been diagnosed in... different medical centres' it might have taken even longer for the dangers of DES to be properly recognised. The case of DES shows that 'the absence of visible and immediate teratogenic effects cannot be taken as proof of the absence of serious reproductive toxicity'.

The fact that DES cost little to make and was not patented, motivated many different companies to produce it, contributing to its rapid worldwide spread. None of these companies drew attention to existing evidence of hazards or problems and none carried out any independent research of their own.

> Pharmaceutical retailers and advertising promoted the effectiveness and safety of DES to doctors and consumers... In fact, some manufacturers promoted it as a panacea for use in all pregnancies. (Harremoes et al., 2002, pp. 96–7)

As Ibarreta and Swan (Harremoes et al., 2002, pp. 97–8) point out, the DES story demonstrates the radical failure of regulators, including the FDA, to properly protect the public. In some European countries, it took another 12 years to withdraw the product after the publication of Herbst's research. DES continued to be used for purposes other than the prevention of miscarriage years after the evidence of its adverse effects. This included use as a morning-after pill for women and growth promoter in animals, from the 1950s to 1979.

Psychotropics and addiction

As Chetley argues,

> the early history of drugs to treat anxiety and sleep disorders set out a pattern of addiction that continues today... one drug after another was introduced as a safe replacement for an addictive drug, only to prove in time [also] to be addictive. First it was alcohol and opium, then morphine, cocaine and heroin; then chloral, bromides and barbiturates; through into benzodiazepines and their more recent replacements – [the serotonin re-uptake inhibitors]. (1996, p. 303)

Starting in 1903, with Bayer's Veronal, barbiturates were originally prescribed as safe and effective 'sleeping tablets'. After 10 years of widespread consumption, 'reports of fatal overdose emerged' (Chetley, 1996, p 303). Solid evidence of their potently addictive character failed to stop over-prescription throughout the 1950s.

> In the UK barbiturate consumption more than doubled in the 1950s and continued to rise into the next decade. A 1962 editorial in *The Practitioner* suggested that too many doctors were taking the line of least resistance and prescribing barbiturates as a blunderbuss remedy for the anxieties and stresses of life. (Chetley, 1996, p. 303)

Chetley (1996, p. 305) points out that doctors recognise few valid uses for barbiturates today, apart from supplying elderly addicts, and as anticonvulsants, despite their continued availability in many countries. They remain 'drugs of choice for suicide'.

Benzodiazepines were the next generation of widely marketed and prescribed treatments for anxiety and sleep disorders during the 1950s and 1960s: Librium and later Valium and Mogadon, were again 'launched with extensive promotion that focused on the safety of the drug compared to barbiturates, that failed to mention the possibility of dependence' (Chetley, 1996, p. 305), and attempted to highlight significant differences based on 'poorly controlled trials'. As Chetley (1996, p. 306) points out, these drugs, again, were far from safe. 'Between 15 and 44% of long term users became dependant on the drugs.' They masked, rather than cured, anxiety, which returned when the patient went off the drug. Overdose remained a problem, with huge numbers of victims. Common side effects included drowsiness and lack of coordination, and long-term use caused psychological impairment and brain damage.

More recently, the serotonin reuptake inhibitor Paxil (known as Aropax in Australia) was approved for treatment of 'social anxiety disorder' as well as depression. There are serious issues of whether any such condition actually exists or whether it should be treated with a powerful antidepressant drug. There are also serious questions hanging over this whole class of drugs. Bristol Myers Squibb's Serzone was withdrawn following 'evidence linking it to hepatitis and liver failure' (Moynihan and Cassels, 2005, p. 33). Twenty-five per cent of those trying to get off Paxil experience 'worrisome withdrawal symptoms' (Moynihan and Cassels, 2005, p. 33). Sexual difficulties are associated with Prozac, Paxil and Zoloft, along with an increased risk of suicidal thoughts and behaviour among children and adolescents.

Under pressure from consumer activists and others, the FDA's re-examination of all of the company trials on children and adolescents, including the unpublished trials, suggested an increased risk of suicidal thoughts and behaviour from 2% (in control groups) to 4%. Furthermore, for most of these drugs (apart from Prozac) the clinical trials provided no evidence that they worked any better than a placebo in reducing depression in children. In

2003 British authorities moved to try to stop the drugs being prescribed to children, at a time when more than 10 million prescriptions per year were written for people under 18 in the US (Moynihan and Cassels, 2005, p. 34).

Overuse of antibiotics

In the 1950s and 1960s (and right through to the 1980s and beyond) antibiotics were widely mis-prescribed around the developed world. Powerful antibiotics were widely used to treat minor infections, including tonsillitis, or to treat the common cold, which is caused by a virus. A third or more of hospital patients were routinely – and frequently inappropriately – prescribed antibiotics. It has been argued that in the developing world at the time most antibiotic use was inappropriate, with medications 'available without prescription used for too short a period, at too low a dose or without proper indication' (Chetley, 1996, p. 87). Many anti-diarrhoeal preparations containing antibiotics were on developing world markets, even though, according to the World Health Organization (WHO), 'antibiotics are not effective against most organisms that cause diarrhoea. They rarely help and can make some people sicker in the longer term' (Chetley, 1996, p. 87).

Back in 1987, an international task force studying antibiotic resistance, noted that

> although antibiotics have saved and improved more lives than any other class of medicines, their use has set in motion the biggest intervention in population genetics seen to date on this planet. The effects of that intervention are seen in the distributions of antibiotic resistance genes throughout the world's bacterial populations. (Chetley, 1996, p. 79)

WHO investigations found 'a correlation between the occurrence of multi-resistant bacteria and antibiotic consumption patterns in 12 countries' (Chetley, 1996, p. 81). Investigators concluded that 'the increased frequency and spread of resistant bacterial strains is a consequence of the inappropriate use of antibiotics in outpatient as well as inpatient care' (Chetley, 1996, p. 81).

As we have seen, such overuse – and misuse – has continued through subsequent decades to the present day. In particular, we have seen the continuing massive misuse of antibiotics in intensive industrial farming, to the value of US$20 billion per year, with livestock in the US consuming 'eight times more antibiotics by volume than the human population' (Weis, 2007 p. 72). Such antibiotics have been used in sub-therapeutic doses to enhance growth. In other cases they are necessary to combat disease amongst animals suffering appalling stress, overcrowding and compromised immune systems. This can lead to antibiotic resistance in bacteria, particularly *Staphylococcus*, undermining the effectiveness of antibiotics in treating humans.

Thalidomide

Unlike DES, psychotropics, antibiotics and many other disasters, which caused comparatively little public disquiet in the 1960s, the thalidomide disaster was a major setback for the industry. This powerful teratogen and neurotoxin was prescribed to pregnant women to treat nausea. Despite clear evidence of toxicity in animals, an obstetrician was persuaded to sign off on a paper, put together by the Richardson Merrill company, recommending the drug for such nausea treatment, and human trials were started. US doctors continued to run 'clinical trials' for months after clear evidence of major harm to human foetuses in Europe and Australia. When the FDA finally sent investigators to recall the drug from the offices of doctors supposedly running the trial, they found that most had not kept records of how much they had received or handed out (Shah, 2006, p. 40).

This disaster motivated new legislation, with the 1962 amendments to the *Food and Drug Act* requiring much quicker recall of dangerous drugs. New drugs were now required to prove themselves both safe and effective before being allowed on the market; through first testing on animals then through randomised placebo controlled trials in humans.

> Drugs approved between 1938 and 1962 would have to submit evidence retroactively showing that they worked or risk being banned from the market.
> (Shah, 2006, p. 41)

The National Research Council (NRC), given the job of testing already approved drugs, pulled 300 of them off the market as unsafe and/or ineffective.

These developments restored public confidence in the system. This was further boosted by wide publicity for new genetic engineering techniques from the early 1970s. A notable early success in this area was Herbert Boyer's isolation of human genes directing insulin production and transfer of these genes into bacteria to allow the mass production of human insulin.

Privatisation and intellectual property

A new phase of rapid growth in the industry was initiated by Ronald Reagan's neoliberal political and economic revolution of the early 1980s.

At that time legislation was pushed through the US Congress, encouraging universities and small businesses to patent discoveries coming from publicly funded research (via the National Institute of Health) and grant exclusive licences to drug companies to manufacture and market such products. After this, big drug companies came increasingly to rely upon publicly funded academics and small biotechnology companies to perform the research for new drugs, rather than funding their own research. Researchers in universities and hospitals came to direct their efforts towards financial gain rather than objective knowledge or pressing need. This has increasingly led to the creation of drugs for everyday consumption by richer, generally older, healthier people – drugs to address depression, anxiety, impotence, heartburn, high cholesterol, high blood pressure and obesity.

Another series of laws extended monopoly rights for brand named drugs. Company lawyers, lobbyists and pro Big Pharma (the major pharmaceutical companies) politicians built upon and further extended these laws to increase the effective patent life of such drugs from eight years in 1980 to 14 years by 2000. As patents expire, generic versions typically sell for 20% of the cost of the brand names. Extra years of patent protection can mean billions of dollars of extra revenue for the big corporations (Angell, 2005, p. 9).

Neoliberals and corporate interests argue that it is both just and efficient to treat knowledge as private property through protection of intellectual property rights. This ensures a 'just reward' for past intellectual labour, and an incentive to encourage more socially useful creative research endeavour in the future. By rewarding those with a track record of such useful research in the past, it provides them with the means to finance further research.

Such rights are currently divided into a number of groups including patents, copyrights, trademarks, and industrial designs. Patents protect knowledge that is applicable to industry or other economic activities. An idea is considered worthy of patent if it is new, not obvious, and of practical use.

As noted Chapters 2 and 3, the World Trade Organization (WTO) provides the institutional framework for the main commercially orientated international treaty on patents – the *Agreement on Trade-Related Aspects of Intellectual Property Rights* (TRIPS). As the WTO outlines, the aim of the basic patent right is to provide the present owner with the legal means to prevent others from making, using or selling the new invention for a limited period of time, 'subject to a number of exceptions' (De Feyter, 2005, pp. 177–8). The limitations and exceptions are supposed to reflect the need to balance the interests of the producers of the ideas with the interests of users.

Corporate interests have tipped the balance very much in favour of corporate patent holders, pushing for stronger and longer periods of patent protection, with fewer limitations on and exceptions to such protection. This has raised particular issues in relation to pharmaceutical patents.

At the time of formation of the WTO, in 1995, member nations were required to amend their legislation to recognise 20 year patents on drugs. As Angell (2005, p. 206) points out, at this time, many countries did not consider drugs as appropriate objects for patenting. The WTO rules allowed for government issue of 'compulsory licences' to allow generic manufacture of in-patent drugs in cases of 'public health emergency'. Poorer countries were required to comply by 2005; later extended to 2016 for 30 'least developed' countries. Following compliance, the prices of basic drugs around the developing world increased by 200–300% (Usdin, 2007, p. 54). De Feyter (2005, p. 178) highlights the role of the patent system in directing research towards 'profitable diseases, or conditions that can profitably be marketed as diseases', and a call by the High Commissioner for Human Rights for alternatives to TRIPS to provide incentives for 'research into diseases that predominantly affect poor people' to 'enable poorer states to fulfil their obligations under the right to health'.

Another issue is the patenting, by transnational companies and public authorities, of traditional knowledge, held by local societies or Indigenous people, of medicinal or nutritional effects of natural resources. In 1993, for example, the US Department of Health and Human Services applied for – and later received – a US patent upon a compound – with strongly anti-viral properties – extracted from a South Australian smoke bush, without any reference to the rights and interests of Aboriginal people who had used the smoke bush as a medicine for hundreds of years. In recent years, there has been a shift away from simple theft to co-option, with particular Indigenous groups being bribed to sell their traditional knowledge. Other such groups are not only denied a cut of the profits, but possibly also denied access to traditional remedies.

Probably the most important potential conflict between intellectual property and human rights concerns access to essential drugs. Extended patents allow extended monopoly pricing of potentially life-saving drugs, depriving poorer people of access, and therefore also of their lives.

The Clinton administration threatened trade sanctions in the late 1990s in response to South Africa's plan to import generic versions of in-patent AIDS drugs to fight the epidemic in that country. In the face of massive public anger around the world, the US government backed down and some drug companies dropped the price of some of their AIDS drugs in Africa. However, these drugs remained more expensive than generics made in India.

Later the Bush administration stood alone in opposing relaxation of patent protection in poorer countries. The US originally agreed to let poor countries manufacture their own generics but not import them, but the poorest countries had no way of doing this. In 2003, the General Council of the WTO allowed that the *Doha Declaration* of 2001 could be interpreted to allow compulsory licences to be issued for the export of generic versions of in-patent drugs to countries unable to make such drugs themselves, but only under particular conditions (De Feyter, 2005, p. 180). When the US voted US$15 billion of federal finds for combating HIV in the developing world, they refused to allow any of it to be spent on generics. A third of the money was required to go to programs to promote abstinence until marriage – with no evidence to support their efficacy. Bush later radically cut the funding to the whole program anyway.

In 2004, the WHO estimated that over 1.7 billion people, mainly in the developing world, had inadequate or no access to essential medicines. Millions continue to die each year of HIV, malaria, TB and other tropical infectious diseases because they have no access to medicines which are more or less readily available in the developed world.

Selling sickness

Long-term patent protection offered huge scope for ongoing profitability through clever marketing, more than research. But the wealthier older people of the developed world, and particularly of the US, were becoming healthier, and less in need of medical interventions. As Shah (2006, p. 44) notes, there was a fall of 62% in the rate of deaths from heart disease and a 21% fall in deaths from stroke 'between 1965 and 1996'.

If the pharmaceutical companies were committed to fighting sickness

> they'd have to make do with markets with minimal buying power, from the tuberculosis ridden inner cities of the US to malarial sub-Saharan Africa and tropical Asia. A more lucrative approach, albeit a smaller contribution to public health, was to encourage wealthier, healthier customers to pop pills despite their relative vigour. After all, no FDA regulation requires drug-makers to invent high priority drugs. (Shah, 2006, p. 44)

In their book, *Selling Sickness*, Moynihan and Cassels (2005, p. x) explain how the marketing strategies of the world's biggest drug companies came to 'aggressively target the healthy and the well', aiming to turn 'the worried well' into the 'worried sick'.

They map out the key features of such strategies: little known conditions are given renewed attention; old diseases are redefined and renamed; whole new dysfunctions are created. The ups and downs of daily life become mental disorders, common complaints are transformed into frightening conditions (Moynihan and Cassels, 2005, p. x). Shyness becomes 'social anxiety disorder'. Pre-menstrual tension (PMT) becomes the much more serious sounding 'PM dysphoric disorder' (needing treatment with Zoloft). Distraction becomes 'adult attention deficit hyperactivity disorder (ADHD)'. Baldness, wrinkles and sexual difficulties become serious medical conditions. Healthy middle-aged women are encouraged to think of themselves as suffering from silent bone disease (osteoporosis), healthy middle-aged men from silent heart disease (high cholesterol) (Moynihan and Cassels, 2005, p. x).

According to the industry's paid medical experts, most middle-aged men should be taking Viagra on a daily basis 'to prevent impotence' (Moynihan and Cassels, 2005, p. 183).

> 90% of elderly people in the US now have a condition called high blood pressure [controllable by drugs], about half of all women have a sexual dysfunction called FSD [treatable with Viagra or testosterone patches], and more than 40 million Americans should be taking drugs to lower their cholesterol. (Moynihan and Cassels, 2005, p. xiv)

With help from the media, the labelled conditions are portrayed as widespread, serious and treatable with drugs: 'alternatives are swept away in a frenzy of drug promotion'.

> While the boundaries defining disease are pushed out as widely as they can be... the causes of these supposed epidemics are portrayed as narrowly as possible... a major problem like heart disease can be reduced to a narrow focus on a person's cholesterol level or blood pressure... Preventing hip fractures among the elderly becomes a narrow obsession with the bone density numbers of healthy middle aged women. Personal distress is seen as largely due to a chemical imbalance of serotonin in the brain – an explanation as narrow as it is outdated. (Moynihan and Cassels, 2005, p. xiv)

Drug company money supported research in the 1990s, by Dr Ron Kessler, which suggested that 30% of people in the US in any given year suffer from a mental illness – 10% from 'major depression' (Moynihan and Cassels, 2005, p. 30). This figure was then widely cited around the world, in marketing and elsewhere, helping to 'build an impression of untold millions... undiagnosed and untreated', despite conflicting evidence from other investigators (Moynihan and Cassels, 2005, p. 28). While the theory of serotonin imbalance as a cause of depression has not been verified by subsequent research, it still

> ...makes good marketing copy. As psychiatrist Dr David Healy says, ' it's the kind of thing a GP can use when they're trying to persuade a person to have pills'. (Moynihan and Cassels, 2005, p. 28)

It's also 'the kind of rationale that drug reps... can use to persuade physicians to use their products' (Moynihan and Cassels, 2005, p. 28). It's certainly easier than beginning to think about how capitalism makes people depressed and sick.

As Moynihan and Cassels say, the common feature in all the different promotional strategies is 'the marketing of fear': a process of situating disease 'at the centre of life'.

> The fear of heart attacks was used to sell women the idea that the menopause is a condition requiring hormone replacement. The fear of youth suicide is used to sell parents the idea that even mild depression must be treated with powerful drugs. The fear of an early death is used to sell high cholesterol as something automatically requiring a prescription. (2005, p. xv)

The industry defends its marketing emphasis as 'raising awareness about misunderstood diseases and providing quality information about the latest medicines' (Moynihan and Cassels, 2005, pp. xv–xvi). It is about educating and empowering consumers through advertising.

Certainly, marketing has been seriously successful in encouraging drug consumption, with young Australians' consumption of antidepressants increasing tenfold between 1990 and 2000, and Canadians' consumption of anticholesterol drugs increasing three hundredfold over the same period. It has been particularly successful in the US with 5% of the world's population consuming 'almost 50% of the global market in prescription drugs' (Moynihan and Cassels, 2005, p. xi) and spending increasing 100% in the six years to 2004. Money spent on heart medicine and antidepressants has doubled in less than five years.

Scientific evidence

One of the ways in which doctors have been persuaded to prescribe and patients to consume drugs that they would probably be better off avoiding, is through systematic misrepresentation of the results of drug tests. This has been achieved through publicising relative rather than absolute risk reductions associated with the drugs in question. As Moynihan and Cassels explain

> advertisements to doctors and patients will claim, for example, that a drug offers a 33% reduction in [the] risk of heart attack, without explaining that in actual fact you may have to take the medicine for 5 years [at potentially significant cost in money and side effects] in order to lower your risk from 3% to 2%. (2005, p. 86)

This is particularly significant for drugs that treat high blood pressure and high cholesterol as risk factors for heart attack and stroke. In both areas, trials indicate that drugs produce only very modest risk reduction for individuals who have no other major risk factors (of prior heart attack, angina, diabetes, smoking, being aged 65+). As Moynihan and Cassels (2005, p. 86) point out, 'the risk of a non-smoking [non-diabetic] 65 year old man [in North America] having his first heart attack over 5 years is about 5 to 6%'. Drugs can reduce this by 1–2% but so can diet and exercise.

Gigerenzer (2002, p. 34) cites the case of the West of Scotland Coronary Prevention Study examining the effects of the anti-cholesterol drug prevastatin sodium. A press release declared that the study had established that 'People with high cholesterol can rapidly reduce... their risk of death by 22% by taking [that already widely] prescribed drug'. As Gigerenzer says, 'studies indicate that a majority of people think [this means that] out of 1000 people with high cholesterol 220 of these people can be prevented from becoming heart attack victims [through the use of this drug]'. In fact, the clinical trial showed that out of 1000 people taking prevastatin over five years, 32 died compared to 41 in a placebo control group.

The relative risk reduction was indeed 22% and the drug manufacturers were happy to use this figure in promoting their product. However, the absolute risk reduction, probably of more interest to both potential users and governments subsidising public medicine, is 0.9%. 'Prevastatin reduces the number of people who die from 41 to 32 in 1000. That is, the absolute risk reduction is 9 in 1000' (Gigerenzer, 2002, p. 35).

Furthermore, the number of people who must participate in the treatment to save one life is 111, because nine in 1000 deaths, which is one in 111, are prevented by the drug. This could be a substantial cost to a public health system, with the money better spent elsewhere. Most obviously, there is the possibility of much cheaper, less dangerous non-drug based means of reducing high cholesterol.

Drug companies and the Free Trade Agreement

As a result of these developments, the big drug companies have been amongst the most profitable business operations in the world for more than 20 years. As Marcia Angell (2005, p. xxiv) points out, in the US the drug industry ranked as the most profitable from the early 1980s, falling to third on the Fortune 500 list only in 2003 (behind mining and banking), with prescription drug sales continuing to grow (pushing up overall health costs at a rapid rate) and now valued at US $200 billion a year (not including drugs administered in hospitals, nursing homes and doctors' offices.) Weiss, Thurbon and Matthews (2004, p. 71) state that in 2002 the top 10 drug companies made profits of US$35.9 billion, 'maintaining this level when the rest of the US economy suffered a downturn and amounting to more than half of the combined profits of the Fortune 500 companies in that year'. The top 10 drug companies had profits of 25% of sales in 1990 and 18.5% in 2001 (compared to 3.3% for other industries in the Fortune 500 at the time). More people are using more, and increasingly expensive, drugs; while escalating numbers of people cannot afford to pay for the drugs they are prescribed. A similar amount is spent around the rest of the world (Angell, 2005, pp. 10–11).

The extent of drug company profits gives the companies huge power to influence the political process. Economist Joseph Stiglitz (2006, p. 191) notes that the pharmaceutical industry 'ranks top in terms of lobbying money and the number of lobbyists employed (3000) to influence US Government policy'. He highlights the US $759 million spent by pharmaceutical companies between 1998 and 2004 to influence 1400 bills in the US Congress their success reflects their investment... the US government has made their interests paramount in international trade negotiations, and under the new Medicare drug benefit the government is proscribed from bargaining for lower prices – a provision worth billions of dollars just by itself. (Stiglitz, 2006, p. 191)

As Weiss, Thurbon and Matthews (2004, p. 59) observe, the big pharmaceutical companies 'are not just wealthy, they are dangerously wealthy', their ruthless 'use of funds to pursue their own political agenda' extends not only to the shaping of the WTO rules but also to the shaping of specific bilateral trade negotiations by the US government. It seems that a major aim of the Free Trade Agreement between the US and Australia, from the US perspective, was to undermine the Australian Government's Pharmaceutical Benefits Scheme (PBS) which has ensured universal access to 'lifesaving and life-improving drugs' for all Australians.

The PBS, established under the *National Health Act* (NHA) of 1953, acts as a

> public wholesaler of drugs equipped with compulsory purchase powers. It deals with the international pharmaceutical companies as a strong purchaser, with powers to list drugs or not to list them. If they are listed, at a price negotiated through the PBS, then Australians have access to these drugs at subsidised prices. If not... they can be supplied, on prescription, at non-subsidised prices. (Weiss, Thurbon and Matthews, 2004, p. 61)

The pharmaceutical companies need to get new drugs approved for safety and efficacy by the Therapeutic Goods Administration (TGA). A committee called the Pharmaceutical Benefits Advisory Committee (PBAC) decides whether the drug is effective and cost effective enough to be listed, and another called the Pharmaceutical Benefits Pricing Authority (PBPA) advises the Department of Health on an appropriate price for bulk purchase of the drug (Weiss, Thurbon and Matthews, 2004, p. 62).

The Pharmaceutical Research and Manufacturers of America (PhRMA) has won government support for destroying the PBS as an 'unfair trade barrier', preventing companies recouping massive research and development (R&D) expenses (Weiss, Thurbon and Matthews, 2004, p. 59). The Free Trade Agreement (FTA) includes a new 'review' process to 're-assess' PBS decisions in light of drug company criticisms, even though they could previously appeal PBAC decisions in the Federal Court. The PBAC is now forced to justify its decisions, while the drug companies keep secret 'true production costs and test results' (Weiss, Thurbon and Matthews, 2004, p. 79). The FTA gives the US a legal basis for acting in support of the drug companies with the threat of trade sanctions against Australia.

The FTA established a new 'Medicines Working Group' with US and Australian membership to make 'innovative medical products available more quickly' (Weiss, Thurbon and Matthews, 2004, p. 64). As Weiss, Thurbon and Matthews say, this group 'has no clear mandate or rationale' in the FTA, but it clearly serves the US as a means of subverting the PBS and subordinating the PBAC and PBPA to US domination, with expensive 'innovative' drugs replacing cheaper and efficacious generics in the future.

Despite vigorous US opposition, the *Doha Declaration on the TRIPS Agreement and Public Health* of 2001 required that WTO agreements 'should be interpreted and implemented in a manner supportive of WTO members' right to protect public health and... promote access to medicines for all' (Weiss, Thurbon and Matthews, 2004, p. 65). There is no reference to any such guaranteed public access in the FTA. The Canadian Government took the Doha principle seriously by enacting legislation to allow generic versions of drugs to be approved by Health Canada 'prior to the patent becoming invalid', using data from the earlier approval of the patented version. Chapter 17 of the FTA has been designed to specifically exclude any such move by the Australian Government (Weiss, Thurbon and Matthews, 2004, pp. 65-7).

As Weiss, Thurbon and Matthews further point out, while the Howard Government tried to assure Australians that the changes to the PBS would make little difference to drug costs and accessibility, US senators, supported by the drug companies, were congratulating the US Trade Representative on ensuring higher prices paid by the PBS for US patented drugs. With anger and opposition to the prohibition of Medicaid negotiating for lower drug prices increasing in the US, and State governments struggling to try to establish PBS style purchasing consortia to bring down drug costs, pressure from Big Pharma has contributed to radically undermining the Australian system, which had previously inspired such reform movements.

Weiss, Thurbon and Matthews (2004, p. 82) ponder the question of what motivated the Howard Government to institute first the removal of all the original PBAC members and then the dismantling of the PBS as dictated by US Big Pharma. To try to answer this question they highlight the large number of staff from the Howard Government 'appointed to rewarding jobs with US drug giants or their industry representatives'.

Drug development

Learning about the nature of a particular disease, including the molecular biochemistry of the condition, is the necessary first step of the development of a genuinely new and effective drug or other treatment. By understanding the chain of events involved in the disease, researchers hope to find a specific link they can target with a specific chemical intervention. This can be a very lengthy and complex process. Contrary to industry propaganda, the research is almost always carried out at universities or government research laboratories. In the US most of this research is supported by the National Institutes of Health (NIH).

As Weiss Thurbon and Matthews point out, between 1965 and 1992, 15 of the 21 most important drugs developed in the US were based on knowledge and technique from federally funded research.

> In most cases, the research had gone beyond concept stage to the molecular level, which means that the most risk intensive phase of R and D was borne by the public sector. Indeed, most of the R&D costs of seven of the best-selling drugs for cancer, AIDS, hypertension, depression, herpes and anaemia were funded by the NIH. (2004, p. 72)

In Australia, the government's ongoing support of health and medical research is principally directed through the National Health and Medical Research Council (NHMRC). In 2008, approximately A$629 million was spent on 3976 separate awards. In 2009, approximately A$681 million was spent on 4079 awards, including NHMRC grants, funding for prestigious Australia fellowships to support Australia's best researchers, and support for a range of medical research institutions (this latter funding is administered by the Department of Health and Ageing) (http://www.nhmrc.gov.au/grants/partnerships/index.htm).

Once a disease is understood well enough to suggest a possible means to cure, alleviate or prevent it, once an infective agent has been identified, for example, and its mode of action mapped out, then the search is on to discover or to produce a molecule that can safely and effectively address the condition. Drug companies focus on selecting or producing relatively small molecules by chemical means. Biotechnology companies focus on making, modifying or combining very large organic molecules or cells from living biological systems.

The preclinical phase of drug development, which may or may not involve drug companies, involves the use of cell cultures and animal testing to find and study promising agents for treatment. Companies keep big libraries of candidate drug molecules, which can be rapidly screened by computerised methods, to see if they are likely to fulfil the role identified by the basic research. Researchers are continually synthesising or extracting new potential drugs from animal, plant and mineral sources.

Animal testing

Scientific and ethical issues of research involving animals are considered in depth in many research ethics and bioethics texts. Animals play a central role in preclinical trials and basic research because of their physiological and biochemical similarities to humans. In most jurisdictions, animals now have some legal protections. Researchers are now required to gain the approval of animal ethics committees which demand humane treatment of animals in the research process. In Australia researchers using animals have to be licensed by state ministries, which enforce animal welfare acts.

Nonetheless, animal research, and the use of animals to train medical students in surgical procedures, sometimes still includes vivisection or mutilation and can result in serious injury or death to such animals. Utilitarian arguments justify the sacrifice of animals to maximise good consequences for humans in the absence of adequate alternatives for testing the efficacy and or toxicity of particular products or treatments. However, the father of utilitarianism, Jeremy Bentham, clearly included animal pleasure and pain in his utilitarian calculation.

Some critics maintain that animal research has played no significant role in discoveries which have provided major benefits to humans. There have certainly been many cases where animal results have failed to prove relevant to humans because of significant differences in physiology, biochemistry, size, psychology and laboratory conditions. Supporters can point to some cases of clear human benefit, and to the limitations of currently available alternatives such as tissue cultures and computer simulations in some situations.

In recent times, cognitive ethologists have extended neurobiological, behavioural and evolutionary investigations of animal consciousness, intelligence, reasoning, emotions, communication, morality and belief. Results indicate that many different species, including fish and crustaceans, are conscious and feel pain. Many species experience a range of emotions, have reasoning abilities, some can use language, have a self-concept or follow moral rules. In some cases animals behave with more concern for other animals than humans do for other humans. In research that probably would not be allowed today, hungry rats were found to stop pressing a lever for food when they saw that this inflicted pain on other rats, showing their moral superiority to chief executive officers (CEOs) and major shareholders of transnational corporations (TNCs), for example, who have continued to push the pedal to extract the last drops of blood from indigenous mine and plantation workers (Bekoff and Pierce, 2009, p. 96).

According to Peter Singer, if we should not perform an experiment on a human, then we should not perform that experiment on an animal. Those who think otherwise embrace an unjustifiable form of discrimination called species-ism: the view that our species is morally superior to other species and that members of our species deserve to be treated differently.

According to another ethicist, Tom Regan, animals have moral rights based on their interests. Animals have interests in not being killed, harmed or placed in captivity. Since animals have rights, they can only be used in experiments if they can give consent or if we can give consent for them; they should not be conscripted and sacrificed in the service of science. Since animals do not choose to be in research and we can't make choices on their behalf, virtually all animal experimentation should be stopped (Mappes and DeGrazia, 2006, pp. 276–98).

Human testing

Phase I of a clinical trail usually involves a small group of normal volunteers to establish safe dosage levels and study the drug metabolism and side effects. The exceptions are cancer and AIDS drugs, which are tested on people with the disease even in Phase I. Promising candidates move on to Phase II. This involves as many as a few hundred patients with the relevant disease or medical condition. The drug is given in various doses and the effects normally compared with a placebo control group (Angell, 2005, p. 27).

With promising results in Phase II, Phase III clinical trials are instituted, evaluating the safety and effectiveness of the drug in much larger groups of patients, hundreds or tens of thousands, again usually with a comparison group of patients given a placebo or pre-existing treatment (Angell, 2005, p. 27).

Drug companies usually get a patent before clinical testing begins. As such testing is part of the 20 year term of the patent, the drug companies are in a great rush to finish testing and put the drug on sale as quickly as possible. The testing is done by doctors in teaching hospitals and private offices, using their own patients or volunteers (Angell, 2005, p. 27).

To speed up their access to human subjects, the drug companies have encouraged the development of for-profit contract research organisations (CROs), set up to organise and carry out trials. The CROs supervise networks of doctors who are paid to administer the studies and monitor the results. In 2001 such organisations had revenues of US$7 billion paid by the drug companies (Angell, 2005, p. 27).

There is a huge demand for research subjects, with CROs and drug companies advertising widely. They set up patient advocacy groups and disguise recruitment as public service announcements to attract new subjects. Subjects are paid a few hundred dollars for participation, but doctors get thousands for each patient recruited. As Angell (2005, p. 27) says, 'one risk of this... system is that it can induce doctors to enrol patients who are not really eligible...', leading to unreliable experimental results. As she says, 'this is probably often the case'.

A drug company must file a new drug application with the FDA before trials begin, describing the proposed research in detail. It must then file another application to gain approval to market the drug. Advisory committees of outside experts assist the FDA in reviewing this latter application, examining the results of the clinical trials and other supporting evidence. However, it has become increasingly difficult for the FDA to find 'outside experts' without financial interests in the drugs in question – or competing drugs – to provide genuinely objective advice.

Human research ethics

Ethical issues of research with human subjects are central topics of consideration in orthodox bioethics and research ethics texts. This typically includes detailed consideration of past atrocities, including the deliberate infection of impoverished Parisians with infectious diseases by eighteenth century medical researchers, the crimes of the Nazi death camps and Japanese treatment of Chinese prisoners and Chinese villagers, and the extensive use of prisoners, mental hospital patients and soldiers by US government agencies (from the 1930s to the 1960s), including the US Public Health Service (Shah, 2006, pp. 62–5). Many such texts include reference to the US Department of Health's treatment of the poor rural African Americans of Tuskegee (from 1932–72), Jonas Salk and Koprowski's vaccine tests on mentally disabled children in Pennsylvania and New York (and children in Africa) and the Atomic Energy Commission's radioactive contamination of pregnant women, patients, children (in special schools), military personnel and large civilian populations from the 1940s to the 1970s (Resnik, 1998, pp. 135–8).

As Sonia Shah (2006, pp. 69–72) points out, it was not so much the Nazis' medical crimes that prompted the formulation of the Nuremberg Code of 1946 by Andrew Ivy of the University of Illinois and a panel of his colleagues, as the revelations (at the International Military Tribunal set up to try a handful of the Nazi doctors) of similar atrocities by US researchers in government sponsored malaria experiments with prisoners. The US prosecutors wanted to be able to appeal to objective ethical principles to measure the culpability of the accused, and thereby shift attention away from similar atrocities in the US.

The key tenets of the code are summarised by David Resnik as follows:
1. Informed consent – 'human subjects can participate in research only if they give their voluntary informed consent' (Resnik, 1998, pp. 133–4), without any kind of coercion. This was seen at the time as absolutely fundamental and is seen to remain so. Today it is taken to include provision of information about the purposes, methods, demands, risks, potential benefits and possible alternatives at the participants' level of comprehension. It depends on good communication between researchers and participants and it continues to raise numerous and difficult problems.
2. Social value – 'experiments should be expected to yield fruitful results for society' (Resnik, 1998, pp. 133–4).

3. Scientific validity – 'experiments should be scientifically valid and well-designed. Experiments should only be conducted by well qualified scientists' (Resnik, 1998, pp. 133–4) There should be an appropriate congruence between the research design and the aims of the proposal. Today this includes a requirement for peer review and dissemination of the research results.
4. Beneficence or non-maleficence – 'no experiments should be conducted that are likely to result in death or disabling injury. Experimenters should take steps to minimise risks of harm or discomfort to participants' (Resnik, 1998, pp. 133–4). Greater risks might be justified by potentially greater benefits, but participants must be protected from pain and harm.
5. Termination – 'during the course of the experiment the subject may stop participation for any reason; the experimenters must be prepared to stop the experiment if continuation of the experiment is likely to result in injury or death' (Resnik, 1998, pp. 133–4).

There has been a lot written about the failure of the code to preclude risky and dangerous experiments on powerless subjects, without voluntary and informed consent, throughout the later 1940s, 1950s and 1960s. New public revelations particularly by Henry Beecher in 1966, of such widespread abuses (including deliberate infection of patients with cancer cells and intellectually disabled children with hepatitis viruses), of the Tuskegee syphilis investigation up to 1972 and of the Atomic Energy Commission (AEC) feeding radioactive materials to children in special schools up to 1973, prompted the formation of a national commission looking at new ethical guidelines for human experimentation in 1974. In 1975, the US, along with 34 other countries, signed the *Declaration of Helsinki*, produced by the World Medical Association (WMA), a group representing dozens of national physicians' organisations around the world.

This new code emphasised voluntary consent, the use of independent ethics committees and the prioritisation of subjects' well-being above all other concerns. In the interests of justice, research subjects should be assured of access to the best health interventions identified in the study. Following Helsinki, and other discussions of the ethics of human experimentation, we can identify a number of principles, additional to those of the Nuremberg Code, now widely accepted around the world

Respect for the autonomy, dignity, welfare and well-being of participants; taken to include consideration for:
1. Privacy and confidentiality of subjects – experimenters should collect and use personal information only with the consent of the participants and only for specific purposes associated with the experiment. Similarly such information should be disclosed only with the participant's consent, or in the vital interests of others.

2. Consideration for cultural sensitivities of participants and communities.
3. Vulnerable populations – 'experimenters should take special precautions to protect subjects for whom informed consent may be compromised e.g. children or adults who are sick, poor, uneducated, incarcerated or mentally disabled' (Resnik, 1998, p. 134).
4. Justice and fairness – 'selection of subjects to participate in all aspects of an experiment should be fair' (Resnik, 1998, p. 134), with no unfair burden on particular groups, a fair distribution of the benefits of participation and fair access to the benefits of the research. The research outcomes should be made available to the participants.
5. Monitoring – 'researchers should continually monitor experiments to determine whether the benefits outweigh the risks, whether the experiment is likely to yield significant knowledge and so on' (Resnik, 1998, p. 134).

Today nearly all public research institutions and many private companies have ethics committees or review boards that review research on human beings. Some researchers see such rules as unnecessarily restrictive or impractical and as obstacles to socially beneficial scientific advance. Sometimes it is seen as necessary to perform research on subjects who cannot give valid consent, such as children or incompetent or unconscious adults. In such situations parents or guardians can give proxy consent. But they always have a duty to act in the best interests of the individuals concerned. Difficulties with applying the doctrine of informed consent also arise where subjects lack the education or judgment to give completely informed consent.

AIDS

By the 1970s the principle of the *Declaration of Helsinki*, that the well-being of the human subject should take precedence over the interests of science, was generally taken to imply that in cases of serious and life threatening illness, new methods should be tested against the best current methods rather than against placebos or inferior treatments. This would ensure that no subjects were left without a significant possibility of effective treatment. However, this interpretation came under serious challenge with the developing world AIDS epidemic.

In 1984 the cause of AIDS was found to be a retrovirus infecting the immune system by taking over pathogen fighting CD4 cells, causing them to cease fighting infection and instead produce 10 billion new viruses per day, including some mutated forms. In 1987 the anti-AIDS drug Retrovir (AZT) was released by a company later to become part of GlaxoSmithKline. This compound, which was supposed to make the virus ineffective by incorporation into its RNA, was developed through government funding in 1964. Nonetheless, as Shah (2006, p. 79) points out, 'the drug manufacturer decided to charge poorly insured and dying AIDS patients 8000 US dollars for a year's worth of treatment'. Despite later price reductions, as the only approved treatment, AZT was bringing in US$300 million a year by 1994.

Such approval came before any solid experimental demonstration of the efficacy of the drug. The first such solid evidence came from a trial testing whether AZT could prevent pregnant HIV positive women from infecting their babies

> In placebo controlled trials in the US and France, AZT . . . [reduced] the transmission of HIV from mothers to children from 24.9% on a placebo to just 7.9%. In the study 100mgs of AZT were administered 5 times a day to infected pregnant women for months before delivery; during delivery the women got an IV infusion of the drug and the baby [received] AZT . . . during its first 6 weeks . . . The entire regime cost around US$800. (Shah, 2006, p. 80)

At the end of 1995, the FDA approved new anti-HIV drugs called protease inhibitors, which disable the virus by preventing it from reproducing within immune cells. Months later, a third new class of drugs became available, non-nucleotide reverse transcriptase inhibitors, which block the virus' RNA from converting into DNA, thus preventing it from taking over immune cells. Treating HIV with all three anti-retroviral drugs turned out to effectively suppress the development of the disease, but the drug combination would cost US$15 000 per year.

As Shah (2006, p. 82) says, the question for many people now became that of how to ensure universal access to the new treatments – including treatment for those developing world populations where the disease was most widespread. But some leading medical researchers – including Robert Gallo – argued that the complex multi-drug treatment would inevitably be misused in less developed countries, particularly Africa, leading to multi-drug resistance. The WHO also decided that it was not practical to apply the costly and complex AZT treatment to pregnant HIV positive women around the developing world (passing on the virus to half a million infants each year) and called for the development of 'simpler and less costly' therapies.

One such simpler and less costly possibility was an HIV vaccine. The 'renegade' French scientist Daniel Zagury 'injected an experimental HIV vaccine into healthy children in Zaire even before establishing whether it would harm animals, arguing . . . that conditions in Zaire were so bad that any risk was worth taking . . . ' (Shah, 2006, p. 85). Another was to trial reduced level AZT treatments of pregnant women. By the later 1990s, 16 different trials were running in developing countries, testing reduced levels of AZT or alternative, cheaper, simpler drug therapies to block mother to child HIV infection.

Only one of these trials, that of Harvard researcher Marc Lallemant in Thailand, ran experimental groups on short courses of AZT against control groups receiving the established long course AZT treatment. All the others ran their experimental therapies against placebos, with experimenters arguing that placebos 'were no worse for HIV infected women in poor countries than what they would normally encounter, that is, no treatment at all'. As Shah (2006, p. 91) points out, 'this meant that western scientists were allowing hundreds of HIV infected pregnant women in their care to deliver their babies unprotected'.

In 1997, a number of doctors and ethicists attacked such placebo controlled trials as totally unethical and against the *Declaration of Helsinki*. Other top AIDS researchers and bioethicists responded by arguing that such critics were themselves unethical and unprofessional

> researchers should provide only the highest standard of care practically attainable in the host country, not those attainable elsewhere in the world, UNAIDS agreed in a guideline document issued the following year. (Shah, 2006, p. 95)

It turned out that the short course of AZT – for around three weeks – worked nearly as well as the much more costly long course. However, it was still too costly for thousands of HIV infected women around the developing world.

In 2006, 4.7 million Africans were in immediate need of life-saving AIDS drugs and only 500 000 had any access to such drugs – 6600 Africans were dying every day from AIDS (Greene, 2006, p. 25). UNICEF reported that a child died or was orphaned by AIDS in Zimbabwe every 20 minutes. Life expectancy at birth in Zambia fell from 60 in 1990 to 36 in 2004. In 2004 in Botswana, 75% of all deaths were attributed to AIDS and a teenager had a more than 75% lifetime chance of contracting the disease.

There are contradictory reports on the success of treatment programs. De Waal (2006, p. 112) refers to hospitals with 'cash and carry' windows, where drugs, including anti-retrovirals, are dispensed, and disturbing stories of anti-retrovirals being retailed with minimum medical supervision or shared with other HIV positive family members without prescription or monitoring. On the other hand, Greene (2006, p. 25) refers to evidence suggesting that higher success rates of patients sticking to strict (AIDS) drug regimes – of around 90% – had been achieved in a number of parts of Africa by 2006. 'In Uganda and Senegal public education campaigns... and drug treatments are beating back the epidemic.'

De Waal (2006, pp. 113–14) points out that only in the richest and 'best-governed' states of South Africa, Botswana and Namibia is there any likelihood of universal public treatment provision. Even in these states, such provision as yet remains disorganised and inadequate.

> Some private sector employers provide ART [anti-retroviral treatment] to their staff, perhaps after restructuring so that low-paid tasks are now contracted out... Better-off individuals... obtain treatment from private clinics. Government institutions – especially the army... provide for their own. NGOs do likewise, often with the support of their donors... For the majority, an uneven array of public clinics, NGO services and churches serves as an ART delivery system... This inequitable system is being established with little public debate and less political outcry. (De Waal, 2006, pp. 113–14)

UNAIDS estimated a cost of around US$20 billion per year to control the pandemic in 2007. However, the world's richest countries contributed less than US$5 billion to fight global AIDS in 2003. By 2005 global AIDS funding had grown to US$8 billion, with about 80% of Africa's HIV/AIDS programs financed from international sources. It is worth comparing these figures with Joseph Stiglitz's and Linda Bilmes' (2008) estimate of the full cost of the Iraq conflict to the US at US$3 trillion.

Drug prices

Drug companies argue that high drug prices (around US$1500 per annum on average for the most widely used prescription drugs in the US) are necessary in order to fund a huge R&D budget required to develop new and improved medicines. They have touted the figure of US$900 million to get a new drug to market. But their own figures indicate that

> ... in 2002 the ten US drug companies in the Fortune 500 list had combined worldwide sales of about $217 billion and spent just over 14% of that on R & D, they had a profit margin of 17%... [and] spent 31% of sales (about US$67 billion) on marketing and administration (Angell, 2005, p. 48).... In 2004, Pfizer, the largest drug company, had a profit margin of nearly 22% of sales (which were US$53 billion). The same year, it spent 32% of sales on marketing and administration and only 15% on R&D. Altogether, the nine US drug companies listed in the Fortune 500 had a median profit margin of 16% of sales in 2004, compared to just over 5% for all the industries listed. (Angell, 2005, p. xxi)

As Angell points out, this R&D budget includes Phase IV studies of drugs already on the market which are often really marketing exercises, rather than research, paying clinicians to use the drugs to introduce them to doctors and patients. R&D expenses are fully tax deductible (at 35%), with drug companies getting additional tax credits (for R&D) worth billions of dollars, so that between 1993 and 1996 they were taxed at 16.2% compared to an average of 27.3% for all the other major industries.

At the same time, the number and quality of new drugs has been declining in recent years as drug companies have claimed that research costs have increased. Most new drugs are not really new at all, but actually just variants on existing drugs, with the originals typically deriving from publicly funded research. As Angell points out, in the five years from 1998 to 2002, only 14% of the drugs approved by the FDA were truly innovative. Nine per cent were old drugs altered in some way and the remaining 77% were classified by the FDA as no better than drugs already on the market to treat the same conditions

> Drug companies have to show the FDA only that new drugs are 'effective' . . . not that they are more effective than or even as effective as what is already used for the same condition. (2005, p. 75)

Weiss, Thurbon and Matthews (2004, p. 72) quote the French drug bulletin *La Revue Prescrire* to the effect that, over a 21 year period, 1780 of 2603 new drugs were 'superfluous products that added nothing to the clinical possibilities offered by already available products'.

> other studies have concluded that . . . of the 445 new patented drugs introduced into Canada from 1996 to 2000, 25 [just over 5%] were major therapeutic improvements, 204 were line extensions [typically a new strength of an existing medication] and 226 represented little or no improvement over existing medicines. (Angell, 2005, p. 75)

Slight molecular variations are recognised for patent purposes as new drugs, even when they have minimal or no functional significance. So companies simply tinker with their own or other companies' existing drugs making minor molecular changes to create large numbers of apparently but not really new products. Sterckx (2005, p. 89) claims that, between 1989 and 2000, 76% of new drugs approved were 'me too' drugs – copies of existing drugs.

As companies can get new patents for old drugs, there is reduced pressure to produce genuinely new drugs. As Weiss, Thurbon and Matthews (2004, p. 73) say, 'the most striking feature of Big Pharma is the paucity of innovation'.

Drug testing

With few public funds available to finance clinical trials, it has increasingly been the drug companies themselves that have overseen the testing of their own products. Regulators have had to rely on the information submitted to them by the drug producers when deciding which products to licence and under what conditions.

It is frequently said that it doesn't matter what sorts of economic interests scientists may have or who exactly funds or organises their research. 'The universally held norms of scientific enquiry in pursuit of the truth make other relationships irrelevant...' (Krimsky, 2004, p. 141). Any scientist who violates these norms to satisfy a sponsor would inevitably be found out and ostracised by the profession. Particular scientists will have favoured theories and approaches but these are subject to ongoing public debate and discussion and ultimately the objective experimental results will determine which theories and approaches are fruitful and which are not (Krimsky, 2004, p. 141).

In fact, there is substantial evidence indicating that financial conflict of interest is associated with significant bias in the construction of experiments and interpretation of results. Surveys have found that industry sponsored research is nearly four times as likely to produce results favourable to the product as NIH sponsored research. Research published in 1986 examined 107 controlled clinical trials comparing new and older therapies. 'Of the papers favouring new therapies 43% were supported by drug companies, compared to 13% favouring the older therapies' (Krimsky, 2004, p. 146). Research published in 1998 found that 96% of researchers publishing articles supportive of calcium channel blockers (CCBs) to treat hypertension had financial relationships with manufacturers, compared to 37% of those publishing critical articles (Krimsky, 2004, p. 148).

It's true, as Krimsky says, that correlations (of for-profit sources of funding and results that favour a new therapy offered by the funding agency) say nothing directly about causation. It's possible that researchers establish financial relations with companies because they have good reasons to believe in the superiority of such companies' products; that companies demand stronger pre-trial assurance of success compared to public authorities before funding trials of particular products. It is also possible that 'clinical investigators, through various subconscious mechanisms that act within the discretionary range afforded to scientists, weight their interpretations of data in the sponsor's favour' (Krimsky, 2004, p. 147).

Bias in this context means finding ways to make it look as if the objective evidence favours a particular result when this is not really the case. This can involve fabrication of experimental results. A recent example here is that of anaesthesiologist Scott Reuben who, over the past 12 years, 'revolutionized the way physicians provide pain relief to patients undergoing orthopaedic surgery'. As reported in *Scientific American*, an investigation has now revealed 'that at least 21 of Reuben's papers were pure fiction, and that the pain drugs he touted in them may have slowed postoperative healing'.

> 'postoperative pain management [of millions] has been affected...,' says Steven Shafer, editor in chief of the journal Anaesthesia & Analgesia, which published 10 of Reuben's... papers. Paul White, another editor at the journal, estimates that Reuben's studies led to the sale of billions of dollars worth of the

potentially dangerous... COX2 inhibitors, Pfizer's Celebrex (celecoxib) and Merck's Vioxx (rofecoxib), for applications whose therapeutic benefits are now in question. (Borrell, 2009)

However, bias isn't always as straightforward as this. It can involve appropriate selection of questions to ask, subjects to sample, quantities or types of substances to administer and results to report, to achieve a particular desired consequence.

Such bias includes researchers extolling the virtues of a drug even though the results don't support such praise. Sometimes only younger or healthier people are enrolled in the trials because they will suffer fewer side effects than older, sicker people prescribed the same drugs. Rather than testing their new products against older, out of patent products, traditional or alternative medicines, the drug companies have gained FDA approval by showing them to be better than placebos, which is all that the FDA requires. When new products are tested against older drugs, the latter can be given in too low a dose or otherwise wrongly administered. Sometimes only part of the data is included in the presentation of the results, for example the first six months of a year long trial, with unfavourable material excluded. Sometimes unfavourable results are simply not reported. Sometimes trials are repeated until chance delivers seemingly favourable results.

They are major problems of 'surrogate end points' and trials too brief to produce meaningful or relevant results. As Shah says

> Instead of having to prove that a new cardiovascular drug reduced mortality from heart disease, e.g., drug companies could show simply that the drug reduced cholesterol levels. Rather than show a new anticancer or AIDS drug extended patients' lives, they could prove instead that the drug shrank tumours or increased white blood cell levels.... (2006, p. 46)

Surrogate end points were hailed by the pro Pharma lobby as reducing the cost and duration of clinical trials. Of course, desperately ill people are willing to pay whatever they can for anything that offers any hope, making surrogate end points massively profitable for the industry

> [But]... numerous studies [show that] drugs that lower cholesterol can increase mortality; drugs that reduce blood pressure [can] increase patients' risks of heart attack; AIDS drugs that increase CD4 counts have no effect on the course of the disease; and drugs that reduce tumours don't extend lives. (Shah, 2006, pp. 46–7)

Moynihan cites the case of the drug flecainide, found in brief clinical trials to reduce irregular heartbeats in people at risk of cardiac arrest. It was assumed that if the drug reduced irregular heartbeats then it would prevent heart attacks, and it was widely prescribed. A belated large-scale trial showed that in fact the drug was causing cardiac arrest in those with less serious symptoms of irregular heartbeat. Health policy researcher Moore estimated that 50 000 patients had died from taking flecainide and the related encainide before they were withdrawn. As Chalmers (Moynihan, 1998, p. 104) says, 'by 1990, more than a decade after these drugs were introduced, it has been estimated that they were killing more Americans every year than died in action in the Vietnam war'.

Krimsky (2004, p. 157) shows how, in some cases, where legal action against the drug company by victims is pending or already in progress, research can be initiated as a 'litigation strategy' deliberately developed to muddy the scientific and legal waters. A case where this became public as a result of the trial discovery process (rather than remaining a secret contractual arrangement between corporations and researchers as is usually the case) concerned silicone breast implant litigation. One of the companies involved hired researchers to do the following:

1. Look at traditional connective tissue diseases and not at the atypical symptoms reported by clinicians in the literature.
2. Include saline as well as silicone implants (which would reduce the cases of concern and the possibility of obtaining a statistically significant finding that silicone caused disease).
3. Use a two tailed significance test considering both the positive and negative impacts of silicone implants.
4. Exclude all women who showed symptoms after 1991 from the study, thereby keeping the mean of 'years after implant' to between seven and nine because some experts believed it took ten or more years for symptoms to develop.

Such research was specifically designed to throw some red herrings into the mix of expert testimony to obscure the true facts rather than reveal them.

Publicly funded research

The FDA requires companies to submit results from all of the clinical trials it has sponsored. It has limited powers – or will – to check out company claims in this area; and does not require companies to publish all such results. As Angell says

> The FDA may approve the drug on the basis of minimal evidence... the agency usually requires simply that the drug work[s] better than a placebo in two clinical trials, even if it doesn't in other trials. But companies publish only the positive results, not the negative ones. Often, in fact, they publish positive results more than once, in slightly different forms in different journals. (2005, pp. 111–12)

Recent drug trials not sponsored by drug companies have shown cheaper, older drugs to be more effective than the much more expensive and more recently developed ones, widely marketed by the industry as significant improvements. The NIH funded and organised ALLHAT (Antihypertensive and Lipid-Lowering Treatment to Prevent Heart Attack Trial) study of high blood pressure treatment, involved 42 000 people over eight years, and compared four types of antihypertensive and lipid-lowering drugs. It found that old style diuretics (on the market for over 50 years) were as good at lowering blood pressure and better at preventing heart disease and strokes than new, best selling and much more expensive calcium channel blockers, alpha-adrenergic blockers and angiostatin-converting-enzyme inhibitors. In fact, the testing of the alpha-adrenergic blocker was cut short because so many receiving it developed heart failure.

Before this trial, the highly advertised newer drugs, made by Pfizer, AstraZeneca and Merck, had largely supplanted diuretics, made mainly by generic manufacturers (with much smaller advertising budgets) and sold for 10 or 20 times less than the new drugs. This is money that could have been used in providing genuinely valuable health care. Even after the results of the ALLHAT study were released, producers of the new drugs continued to claim that such drugs were 'as good as' the old diuretics – with Pfizer, for example, selling 'about US$5 billion worth' of its new generation Norvasc – costing up to 200 times as much as the equivalent diuretic – in the year following the publication of the ALLHAT results.

> According to [a] study published in JAMA, almost a quarter of [the many billions spent around the world on blood pressure drugs] could be saved if physicians stuck to cheaper therapies. A similar study undertaken by publicly funded researchers in Norway [estimated that] the UK could save more than $100 million and the US . . . between $500 million and one billion per year. An Australian study estimated taxpayers could save up to A$100 million a year by using more of the older but equally effective medicines. (Moynihan and Cassels, 2005, pp. 95–6)

This puts the US government's legal commitment not to use its monopsony power to bargain for cheaper drug prices – and the undermining of the Australian PBS – into context. Here we are not even beginning to factor in earlier considerations of massive over-prescription of blood pressure drugs as a result of judgments based on relative rather than absolute risk reduction.

Even publicly funded research is increasingly threatened by the conflicting financial interests of leading scientists. Krimsky (2004, p. 22) cites the case of Dr Richard Eastman, leading diabetes researcher working at an NIH research institute. In this capacity, Dr Eastman played a key role in a nationwide diabetes study testing a range of different diabetes treatments with more than 4000 volunteer patients. Warner-Lambert Pharmaceuticals, producers of a type 2 diabetes drug Rezulin, promised to contribute US$20.3 million towards the costs of the investigation in exchange for intellectual property rights in the discoveries made. Eastman expressed enthusiasm for the drug, supported its inclusion in the trials, and also backed its adoption by the FDA in an accelerated review.

After several months of sales of the drug in 1997, the FDA received reports from GPs of patients taking Rezulin experiencing liver failure. 'By the time the FDA cancelled Rezulin it had been linked to at least 90 liver failures including 63 confirmed deaths and nonfatal organ transplants' (Krimsky, 2004, p. 20).

Eastman, while an employee of the NIH, participating in a number of key deliberations concerning Rezulin, was also a paid consultant to Warner-Lambert, giving talks to physicians on prescribing Rezulin to their patients. These deliberations included decisions to include Rezulin in the national diabetes study (potentially driving huge sales increases) and decisions in 1997 to retain the drug in the study after patient deaths were recorded. It was not withdrawn until June 1998. It turned out that 12 or more of the 22 scientists involved in the NIH diabetes study received research funding or compensation from Warner-Lambert (Krimsky, 2004, pp. 20-2).

Informed consent

As noted earlier, and as should now be clear, there remain many and deep problems associated with the principle of informed consent or choice, despite its supposed centrality in research ethics. As Shah points out, few researchers take any serious steps to verify that their subjects actually do understand what the research really involves. As she says, it wouldn't be difficult to do so

> Medical researchers routinely double check . . . and re-analyse nearly every aspect of clinical trials by means of . . . journal articles, conferences, workshops and lectures . . . But in the area of informed consent, an . . . atmosphere of 'don't ask, don't tell' prevails . . . One survey found that while over half of researchers admitted that it would be a good idea to verify their subjects' understanding of experiments, a scant 16% had ever actually done so in their own trials. (2006, p. 147)

Often 'consent' involves a single meeting between an investigator and a prospective participant, with a brief 'explanation' of the trial and of a consent form, a few questions answered and the form signed.

> It is a brief, legalistic exchange that satisfies sponsors, oversight committees, and regulatory authorities, protecting all concerned from liability. But the evidence suggests it does little to enlighten test subjects about the risks of experimentation. (2006, p. 147)

Informed choice of participation is fraught with issues and problems. There is general agreement that it includes information about risks. Risk involves complex issues of probability, statistics and empirical evidence. The likely risks of the new treatment should presumably be compared to those of existing treatments, other possibilities and no treatment. Potential subjects should know if there are any alternative drug or non-drug based treatments, including non-patented traditional remedies, and how they rate in appropriate clinical trials. They should know if the drug under test is just a 'me too' variant of an existing product.

Certainly, subjects need to know how earlier investigations have provided support for the present trial. They should know if any reputable authorities see any problems with the new drug or the nature of the trial and how to contact those who do. How will random selection of experimental and control groups from an original population of sufferers be assured? Will the control group receive the currently best available treatment that is properly administered? Will success be measured in terms of surrogate end points? If so, what is the justification for this?

Who has actually funded the basic research behind the treatment and who will actually have access to the treatment following successful trials? How much is it likely to sell for and why? What are the financial interests of those actually conducting the trial? How exactly did they select the potential participants and was money involved?

As might be anticipated, there are major issues in respect of Western sponsored trials conducted in developing countries. Shah (2006, p. 152) cites evidence of numerous trials carried out without any meaningful consent, including some resulting in subsequent legal action.

FDA issues

Under intensive lobbying by Big Pharma the US Congress enacted the *Prescription Drug User Fee Act* in 1992, which authorised drug companies to pay user fees to the FDA to expedite approval of new drugs. This process was justified as a response to AIDS and cancer activists pressing for quicker access to potentially life prolonging drugs for seriously ill patients (Noonan and Nazario, 2005). But as Noonan and Nazario point out, the acceleration of the evaluation process was applied to all medicines, making it easier to get approval for ever more patented 'me too' drugs, with an exponential increase in such patents.

Such fees soon accounted for half of the budget of the FDA's drug evaluation centre making the FDA heavily dependent on the industry it regulates. In 2003 such user fees amounted to US$260 million, with industry financed workers contributing more than half of the FDA drug approval staff. At the same time, an extreme pro Big Pharma republican, Mark McClellan, was appointed FDA Commissioner, and called for drug price increases around the world to match those in the US – to facilitate the development of new treatments.

An independent survey by the US Department of Human and Health Services of FDA scientists conducted after these changes found that: 60% didn't believe that the FDA had time to conduct in-depth science based reviews of new drugs, 48% thought that the FDA didn't do enough to monitor and improve its drug assessment process, and 20% said that they 'had felt pressured to approve or recommend approval' of a drug despite reservations about its safety, efficacy, or quality' (Adams, 2005; Noonan and Nazario, 2005).

As Amanda Spake (2004) points out, according to FDA scientists, for a drug to be denied approval, there must be 'complete certainty... beyond a reasonable doubt' that it will cause harm. This creates an environment where, in situations of uncertainty, drugs are approved, rather than not.

Faster approvals mean greater likelihood of dangerous drugs reaching the market. In the decade after the *Prescription Drug User Fee Act* was passed, a record 13 prescription drugs were withdrawn after causing hundreds of deaths. Fast tracking seems to have been a key factor in the Vioxx story. This fast tracked drug for arthritis pain was heavily promoted by the manufacturer, Merck, despite the company's knowledge of serious problems at the time of testing. The drug was withdrawn from sale in September 2004, after being linked to an estimated 40 000 deaths in the US and 300–1000 in Australia. Richard Horton (2004), editor of *The Lancet*, argued that both Merck and the FDA, 'acted out of ruthless, short-sighted and irresponsible self-interest'. Lawsuits arising from this case involve compensation of between US$10 and 15 billion.

The situation is similar in many European countries, and particularly bad in Australia, where companies pay 100% of the costs of the public regulator – the TGA. As Viola Korczak argues, there are major ethical and public health concerns regarding the pharmaceutical industry's influence, with the industry testing its own drugs for 'quality, safety and efficacy', and pushing for more rapid approval processes. In particular she raises questions of whether shorter approval times 'reduce scrutiny of test results' and whether drug company research priorities are 'consistent with community needs'.

> ... pharmaceutical companies have more incentive to develop drugs for chronic illnesses because they will be used for longer time frames. The influence of the pharmaceutical industry, lack of transparency in processes and changes in the funding structure of the TGA over the last 15 years, have significantly reduced the effectiveness and independence of the regulator. (Korczak, 2005)

Waivers

As Krimsky explains, the US FDA is part of the Department of Health and Human Services.

> Under the DHHS regulations (5 CFR 2635) advisory committee members may not participate in matters that are likely to have a 'direct and predictable effect' on the financial interests of a person with whom he or she has a relationship (members of household, close friends or employer). Like other federal agencies the FDA can exercise a waiver provision of the US ethics statutes on conflicts of interest. (2004, p. 95)

A waiver may be issued by the appointing officer of an agency only if a prospective member's financial interest is considered so 'remote' or 'inconsequential' as to be 'unlikely' to affect their judgment, or if the need for their services 'outweighs the potential conflict of interest' (Krimsky, 2004, p. 94).

The *Ethics in Government Act* does not require that any such conflicts of interest be made public. Usually they are not made public. The FDA did do so, up until 1992, when it stopped disclosing the particulars of committee members' conflicts of interest, following a series of public controversies, supposedly to protect their privacy.

An investigation carried out by reporters for a US magazine of 18 expert advisory committees established by the FDA's Center for Drug Evaluation and Research, meeting between January 1998 and June 2000, found that

> at least one . . . member had a financial stake in the topic under review at 146 of the 159 meetings (92%). At 88 of the meetings (50%) at least half of the advisory committee members had financial interests in the product being evaluated. The financial conflicts of interest were more frequent (92% of the members had conflicts) at the 57 meetings at which broader policy issues were discussed; however, at the 102 meetings that dealt with a specific drug application, 33% of the experts had a conflict of interest. (Krimksy, 2004, p. 96)

Krimsky raises the question of whether the FDA is issuing so many waivers – in this case for 50% of all member appearances – 'because the reservoir of independent experts – without ties to the industry – is declining so rapidly'. Whatever the causes, the consequences can be disastrous. Krimsksy (2004, p. 23) cites the case of the RotaShield® (rotavirus) vaccine, approved by the FDA in August 1998 but removed from the market 'within a year, after more than a hundred cases of severe bowel obstruction were reported in children who had received the vaccine'. A government committee found that 'the advisory committees of the FDA and the Centres for Disease Control' assessing the vaccine for approval 'were filled with members who had ties to the vaccine manufacturers' (Krimksy, 2004, p. 24).

According to Lazerou, Pomeranz and Corey (1998), this has all contributed to a situation where the industry is responsible for more than 100 000 deaths every year through the sale and prescription of dangerous drugs. This does not include drugs incorrectly prescribed by doctors or severe reactions to these drugs, which total over two million per year.

Drug promotion

The FDA is also supposed to review drug labelling and advertising for accuracy. However, it has very limited resources for this role with 34 people to review 340 000 advertisements in 2002.

Much of drug company advertising is disguised – including US$11 billion worth of free samples given to doctors in 2001 and US$5.5 billion worth of talks with 88 000 sales representatives. Much advertising and marketing is disguised as education in order to avoid legal constraints and bad publicity. Companies are allowed to promote drugs only for the uses and at the doses which are approved. However, doctors are not bound by these regulations.

If companies can persuade doctors to prescribe drugs for off-label uses, they can bypass FDA rules and increase sales. While it's illegal to pay doctors kickbacks for prescribing particular drugs, they can get away with limitless bribes presented as serving educational or research purposes. The corporations pay 60% of the cost of continuing medical education in the US. Doctors are regularly rewarded with fancy food and holidays for acting as 'consultants' or 'advisors'. Meetings of professional associations are now also partly supported by drug companies.

Advertisements masquerade as newscasts and celebrities spontaneously discuss their health problems on news and entertainment shows. After the FDA ceased to require companies to include full information on side effects in TV advertisements in 1997, companies began to bombard consumers with commercials for their latest drugs, getting them to ask their doctors for the new drugs.

The Obama administration has taken positive steps to begin addressing the issue of off-label prescribing, supported by 'kickback' payments to doctors (in this case, including 'all-expense-paid trips to resorts, and free massages', as well as money payments (Arnst, 2009)), with the recent Department of Justice fine of US$2.3 billion imposed on Pfizer for illegal promotion of a number of drugs for uses not approved by the FDA. This included the promotion of the arthritis drug Bextra, as a treatment for severe pain, despite the FDA specifically refusing to support such use, as well as off-label use of erectile dysfunction drug Viagra, and cholesterol treatment Lipitor. It is the fourth such fine imposed on the company since 2002 but it is much larger than previous ones (Barret, 2009).

The company was also required to report regularly to the US Department of Health and Human Services on its continued compliance with the law in this area. The company's lead counsel, Amy Schulman, was reported in *The Boston Globe* (Lannin, 2009) as saying that this agreement brought 'final closure to significant legal matters' and helped 'to enhance our focus on what we do best – discovering, developing and delivering innovative medicines'.

DISCUSSION TOPICS

1 What are 'problem drugs'?
2 Could or should drug patenting cease to be permitted?
3 How can genuinely informed consent be assured in clinical trials?
4 How can drug approval processes be improved?

What is to be done?

Previous chapters have highlighted a range of obstacles currently denying large numbers of people the quality of life and health which existing resources and technology could relatively easily provide for them. Reasons have been provided for believing that the situation will worsen significantly in the future due to accelerated land loss for subsistence producers, climate change, resource depletion and recession, if current policies continue to be followed.

In ethical terms, as considered in Chapters 1–3, existing arrangements radically fail to achieve the utilitarian outcome of welfare or happiness maximisation for people and other animals or human and animal rights protection which could be achieved on the basis of different sorts of social arrangements.

Earlier chapters have highlighted a number of factors contributing to this situation. The principal focus has been upon neoliberal ideologies and practices of competition, deregulation, privatisation, destruction of social welfare provision and prioritisation of private corporate interest and power as forces contributing to increasing inequality, poverty and disempowerment of the great majority of the world's population. Radical and increasing inequality within and between countries resulting from such policies and practices has been centrally implicated in avoidable mortality and morbidity around the world.

Inequality and disempowerment contribute to adverse health outcomes because the disempowered lack the resources to acquire basic necessities in a competitive market situation, or in a situation of monopoly enforced scarcity. Available resources go instead, and in excess, to the wealthy and powerful minority. At the same time, the experience of disempowerment and devaluation of the victims of radical inequality undermines their individual health through weakening their immunity and their individual capacities for healthy and rational lifestyle choices. Stress and lack of both education and reliable information leaves them prey to dangerous drugs, fast foods, pollution, infections and quack medicines.

Implicit in this critique are proposals for reform – basically turning the neoliberal programme on its head. Privatisation, competitiveness and deregulation must be reversed, in favour of collective, planned, democratic control and collaboration.

Chapter 1 looked at an interpretation of the work of influential moral philosopher, John Rawls. This was based upon the idea that true justice in distribution of social benefits and burdens cannot be imposed from above, but rather integrally involves genuinely voluntary and cooperative participation of all those involved in collective social enterprise, in both the determination and the implementation of basic principles of distributive justice within that enterprise. Chapter 4 provided evidence to support the idea that such cooperation is a necessary condition for good physical and mental health for human beings.

There have been many suggestions about how such a principle would operate in practice. Certainly it requires a major shift away from the secret exercise of power by small elite groups – of executives of transnational corporations (TNCs) – of big banks and investment funds, of big energy and civil engineering businesses, of government agencies substantially subordinated to such private interests (like the Food and Drug Administration (FDA) in the US, as considered in Chapters 5, 6 and 9), of the US military diplomatic establishment and right-wing 'research centres' and 'think tanks'. Rather, such power must be shifted to all of those currently subject to such elite control, to genuinely participatory, collective and fully public decision and action.

Production

Utilitarian arguments are frequently used to try to justify the completely non-participatory, non-voluntary or non-democratic structure of modern corporations, including big transnationals. It is maintained that the hierarchical flow of information from the productive base, via various professional overseers and experts, to central controllers and planners, and back again is the only way to organise large-scale productive undertakings. Larger operations can achieve very significant economies of scale which increase the productivity of labour and limited resources beyond anything achievable on a smaller scale.

But, in some areas, at least, there is a real question of whether such scale economies ethically or practically outweigh the advantages of smaller-scale productive organisation, allowing the whole workforce to come to know and respect one another through day-to-day collaboration, running the business on the basis of democratic consensus.

As argued in Chapter 3, in key areas of productive organisation, particularly those involving transport and power infrastructure, there are very strong arguments for major efficiency gains achievable by allowing only a single monopoly supplier for a particular nation state or region. Strong arguments have also been mounted concerning major problems of private ownership and share market finance of large productive undertakings.

Current arrangements of private ownership and centralised control allow elite groups of executives and major shareholders to exploit both employees and consumers, paying monopsony wages and charging monopoly prices, while suppressing new and sustainable technologies and inflicting huge damage upon the natural world. Such elites utilise substantial social surplus to undermine political democracy. Private banks literally create money as debt to boost their profits and fund irresponsible and dangerous business operations while denying funds to safe and sustainable operations.

Most company finance actually comes from loans, including bond sales, and profits, rather than share markets. The latter rather function as a means for wealthy elites to gain control of social surplus they have played no part in creating, while also claiming ownership of the means of social wealth creation. As Wilkinson and Pickett point out

> A smaller and smaller part of the value of a company is the value of its buildings, equipment and marketable assets. It is, instead, the value of its employees. When companies are bought and sold, what is actually being bought or sold is . . . its staff as a group of people, with their assembled skills and knowledge . . . [Such a] concept is the very opposite of democracy. (2009, p. 249)

While contributing little to increasing the development of the productive forces, share markets contribute much to undermining such development through directing social surplus into the luxury consumption of elite minorities, and periodically generating economic crises which destroy such productive forces, producing unemployment and misery on a large scale.

Public ownership of major social productive forces allows such forces to operate in the public interest, which generally means in the interest of efficient provision of necessary goods available to all, based upon low pollution, sustainable technologies and good wages and conditions for the workforce. Ongoing and informed democratic input of the whole social collectivity can play a central role in establishing and implementing the major goals of the production process. Such informed participation could ensure rapid transition to sustainable, low emission technologies of electricity generation, transport, agriculture and construction (along with large-scale reforestation). It could ensure production and use of adequate social surplus to sustain comprehensive welfare provision (of high quality public health care and medical research, education, housing and reliable information provision for all) at home and assist rapid development of poorer nation states and regions overseas.

Such external democratic control need not preclude significant internal democratisation of the operation of such large-scale monopolies. This can be achieved through an increasing sharing of manual and mental productive tasks, with all employees engaged in both sorts of activities, and limited, if any, wage differences between different grades of workers. So too could managers and directors be chosen (and periodically replaced) by lot, so that they are proportionally representative of the various grades of workers within the organisation, or directly elected by, and answerable to, the whole workforce.

In many cases, smaller business operations can be run as collective and collaborative workers' cooperatives, without significant sacrifice of efficiency. On the contrary, as Wilkinson and Pickett (2009, p. 251) note, the Mondragon employee owned and run cooperatives of the Basque region of Spain 'are twice as profitable as other Spanish firms and have the highest productivity in the country'. They refer to the research of Robert Oakeshott (Wilkinson and Pickett, 2009, pp. 250-1), looking at cooperatives in a number of different areas to support the claim that 'a combination of [worker] ownership and participation does indeed have the potential to improve productivity by reducing the conflict of interest'.

Within the public sector, central planning is responsible for the appropriate investment of funds through direct control of production and the redistribution of surplus. State controlled savings banks, lending to businesses that have become workers cooperatives, offer the potential for investment directed in the public interest through conditional lending while also yielding fair returns to savers.

Continued private ownership and control

Without complete state takeover, big corporations can still be directed to function in the interest of society through various kinds of controls of their operations, including restrictions upon access to funding and movement of capital within and between nations and regions. This requires careful central planning and ongoing monitoring.

To the extent that large private businesses continue to operate transnationally, appropriately transnational institutions of governance, regulation and taxation are needed to control their operations. As Jaques Crossart, Kohonen and Mestrum (Kohonen and Mestrum, 2009, p. 158) point out, threat of disinvestment has pitted states against each other in progressively reducing corporate tax rates, to the point where effective corporations tax rates in the Organisation for Economic Co-operation and Development (OECD) are now far below typically regressive consumption tax rates. There have been various proposals as to how to address this issue. One favoured by many is an international 'unitary tax rate', 'wherever a company is established, it would pay a unitary rate on its profits'.

This would only partially address the issue of transfer pricing and would need to be accompanied by ways of ensuring transparency in the reporting of such profits, including the closing down of all tax havens. This highlights the necessity of including public representatives at the highest levels of governance of the corporations in question – possibly government auditors, as suggested by J. K. Galbraith.

Wealth

Socialisation of private wealth is a crucial step towards the creation of a genuine democracy as it winds back the subversion of democratic institutions and practices. It reduces the power of wealthy individuals and groups to radically restrict the scope of voters' meaningful choice to their own selected candidates through funding of election campaigns, and control of media coverage, through bribery and threat of disinvestment.

The generation and perpetuation of great private wealth has crucially depended upon a lack of democratic control of production, a radical inequality of remuneration (up the corporate hierarchy) and the operation of the share market system. The proposed reforms of these practices and institutions can be expected to reduce such radical inequality in the longer term. The fact that existing wealth holders will strive to very actively oppose any such reforms provides a good reason for addressing radical wealth inequality directly in the shorter term. Certainly, anyone who believes in equality of opportunity should be demanding an end to significant private wealth concentration and inheritance because it completely undermines such equality. This was considered in Chapter 3.

As has frequently been argued, a straightforward and systematic transfer of concentrated wealth to public welfare provision can be achieved by property and death taxes. Crossart, Kohonen and Mestrum (Kohonen and Mestrum, 2009, p. 155, pp. 158–9, p. 159), for example, argue that it is hardly unreasonable to expect the 9500 ultra wealthy individuals who hold around US$11 500 billion in offshore tax havens to hand over 1% of such wealth per year to provide $100 billion to 'contribute to the restoration and maintenance of the planet' and to the 'satisfaction of the basic needs of the planet's poorest people'. As they say, 'It is difficult to see why the world's citizens would not agree to demand such a modest contribution'.

In the interests of justice, human rights and social welfare, a comprehensive property tax should be supported by a ceiling on inheritable wealth. There is also a need for transnational intervention to close down all offshore tax havens, or enforce their compliance in the levying of such taxes, and ensure that receipts get to where they are most needed – to poverty reduction, welfare provision and environmental repair.

If share markets continue to operate, unearned share income (other than workers' superannuation) must be taxed at a rate significantly higher than earned income, rather than being untaxed through dividend imputation. Ways must be found to limit the harm of such private ownership through restricting share owners' capacities to do great economic and social damage through panic selling and leveraged buy-outs. Relevant ownership and voting rights cannot be allowed to override the rights and interests of workers who actually create the wealth of the corporations or those in vital need of their products.

Politics

A major problem for liberal democracy is the lack of meaningful democratic participation within the major political parties themselves, with self selecting ruling elites effectively controlling policy, and little meaningful input from rank and file members. When in power, prime ministers and presidents select cabinet members and other senior officials, including High Court judges. In some cases, such leaders exercise near dictatorial power over the legislative process, through application of party 'discipline'. The Westminster first-past-the-post system perpetuates a tyranny of the majority, whereby minority parties which, nonetheless, have significant support in the community fail to have any representation in the legislative assembly. In part, these things are maintained as they serve the private interests of the wealthy individuals who currently finance the operations of 'mainstream' parties, through reducing meaningful public representation and debate.

Such mainstream – and other – political parties could, in theory, be easily democratised, with leaders and parliamentary candidates chosen by, and answerable to, party members. Proportional representation goes some way towards addressing the tyranny of the majority issue, with effective action dependent upon a negotiated agreement between a potentially wide range of different political groups and interests.

There are fairly obvious steps which could and should be taken to reduce corporate subversion of the political agenda and of voters' ideas – in addition to or in lieu of – the reforms of corporate governance already touched upon. Private contributions to the funds of political organisations other than flat rate subscriptions from members should be outlawed. At the same time, ways must be found to prevent parties in power using public funds to preferentially finance their own campaigning (e.g. under the guise of information provision). Extra funds for political campaigning could be supplied by the state equally to all parties or other organisations or in proportion to members' contributions.

All political organisations need to be given genuinely equal access to significant media of mass communication to get their ideas across to voters. This would undoubtedly require an end to the current concentration of media ownership and very strong legal requirements for genuine balance and fairness in reporting and comment. Ways need to be found to hold public officials to account for promises made earlier, once they are appointed. This would imply absolute transparency of all political decision making and regular scrutiny by elected representatives of the public, with the power to censure, prosecute or remove officials from office (Head and Mann, 2009, pp. 242–4).

Aid to the developing world

Preceding chapters have identified a situation where 800 million people in the developing world suffer chronic malnutrition; 1.2 billion are in extreme poverty, 2.8 billion earn less than US$2 per day. All suffer the disempowerment and adverse health consequences of this inequality and poverty. Many are victims of deprivation of land, which has been taken away from subsistence and handed over to increasingly large-scale capitalist farming of cash crops for export, along with large-scale mining operations, oil fields and hydro-power. In Bolivia, for example, 'by the 1990s, 400 individuals owned 70% of the nation's productive land' (Panitch and Leys, 2007, p. 21). As Rosset (2006, p. 5) says, 'around the world, the poorest of the poor are the landless in rural areas, followed closely by the land-poor, those whose poor quality plots are too small to support a family'.

The World Bank gives 'development assistance' to big plantations of export crops in the developing world. In many cases, Indigenous labourers, who have been deprived of their ancestral lands, are forced to work very long hours for below subsistence wages and are regularly threatened and beaten by bosses and overseers. World Bank aid to these operations is justified on the basis of the alleged efficiency gains of an agricultural system built upon 'large scale, industrialised, subsidised and fossil fuel intensive monoculture' with high levels of 'agro-chemicals and fertilizers, coupled with intensive livestock production' run by big businesses, including TNCs, and involving increasing speculation and long distance transport of agricultural exports.

As well as perpetuating poverty, malnutrition and ill health, this system is poisoning the environment and accelerating global warming with destructive consequences as mapped out in Chapter 7. This system is unsustainable in the long run because of the destruction of land fertility and the rising cost of oil inputs (as considered in Chapter 5). Even in terms of current net output per unit area of land, it is matched or exceeded by small-scale, labour intensive, multi-cropped, organic agriculture.

Developing world poverty is also attributable to the absence of industrial development to provide work for displaced populations and improve the terms of trade for these nations. It is attributable to weak or corrupt governments unable or unwilling to address these issues.

Many developing nations are heavily dependent upon mineral exports to the developed world. Orthodox economics recognises that mineral exporters typically fare worse than other countries in the world market because their currency values are inflated by the high value of such minerals. This makes all of their other exports more expensive to foreigners, so other export industries do not thrive.

But for much of the developing world, the problems are rather those of overseas control of such resources. Ineffective taxation regimes fail to provide adequate returns for the loss of such valuable materials. Large-scale devastation and loss of land, produced by mining operations, deprives local farmers and fishers of their livelihoods. Pollution and poor working conditions in the mines contribute to widespread ill health.

These are not just issues of history, with developing regions and governments inheriting conditions completely beyond their control. We have seen how the governments of Cuba, Costa Rica, Sri Lanka and Kerala have been able to provide work for most of their citizens along with comprehensive high-quality health and welfare at minimum cost, despite continued very low levels of gross domestic product (GDP). This, indeed, shows how corrupt, or effectively subordinated to the interests of the rich and the privileged, at home or abroad, are other governments in territories with comparable or higher GDP, which have radically failed to ensure such full employment and basic welfare provision.

It has been suggested that the facts support the need for strong, rather than democratic, government in the developing world, to address urgent issues of absolute poverty. Increasing democratisation is rather an issue for the developed world, seeking to reduce relative inequality. But given the achievements of, at least partially, morally principled and scientifically informed developing world governments, greater democratic participation in the developing world could ensure such principled decision making on a much broader scale.

As Usdin says

> a strong civil society from grassroots activities and NGOs to academics is essential to hold governments accountable... in order for policies and strategies to work, grassroots participation in both the design and implementation is important as are strong communities that can hold all players to account. (2007, p. 142)

Countries and regions without basic infrastructure need to work closely together with each other and with countries that have the resources to accelerate their development, not in free market exchange, but in planned integration of production and distribution.

This requires recognition of the 'odious' nature of most developing world debt – leading to cancellation – and of the radical imbalances in terms of trade and currency values that allow theft of developing world resources paid for by Westerners. It also calls for significant recompense for interest and resources already – illegally and immorally – appropriated, and for the social damage produced by such appropriation. As Kohonen and Mestrum point out, 'between 1998 and 2003 net transfers from poor to rich countries were US$51 to 132 billion per year' (2009, p. 37).

At the forefront of such significant wealth transfer are the big mining corporations, BHP Billiton, Rio Tinto, Vale Inco, Anglo American, Xstrata, Alcoa, Newport and others, generating huge profits by extracting and exporting copper, gold, diamonds, uranium, thorium, platinum, chromium, nickel, iron, lead and coal from developing world territories. Clearly the wealthy elite of the richer countries are in massive debt, both financial and ecological, to the poorer majority in the developing world.

This is often understood in terms of the need for massively increased donor aid from rich to poor countries. As Usdin points out

> ODA [official development assistance] amounts to approximately US$80 billion per year, but in 2005 ActaAid calculated that only one third of this aid is real aid, targeted at reducing poverty... the aid spending on health by all developed nations, $10 billion... is equivalent to the annual amount Europe spends on ice cream. (2007, p. 135)

Such increased aid can be funded either through effective taxation of existing wealth holders and wealth producers, or it can be achieved through takeover of the primary sources of wealth generation by democratically controlled – national or supernational – political authorities. Such increased aid needs to be targeted to rapid – and sustainable – infrastructure development, centred, as Mehrotra and Delamonica (2007) argue, upon interventions in 'health, nutrition, water and sanitation, fertility control and education', along with sustainable electricity generation, public transport and housing provision.

Land reform, minerals and industrial development

The issue of land loss can be immediately addressed through large-scale land reform, with the big estates broken down and reassigned to the currently landless and dispossessed. Such land reform ideally needs to be an organised process, with central government and international organisations taking the lead in a peaceful, rational and fair redistribution of land to give cooperative groups of family farmers adequate resources for survival and surplus production.

In the longer term, such central government supervision will be vital as 'free market' forces cannot be allowed to drive new processes of land accumulation from below, with poorer farmers losing their land to wealthier ones (again) via debt or other dispossession. As Weis (2007 p. 183) says, this issue can be addressed through 'community or state held property with secure renewable leases for farmers... giving farmers motivation and legal protections to cultivate with the long term health of the land in mind'. The state must protect ownership through preventing land sales or thefts and ensuring adequate recompense for appropriate agricultural surpluses, channelled into sustainable development of other parts of the economy.

As in Cuba, executive authority must assist in ensuring that appropriate technologies and education in their use are provided to farmers to allow for sustainability. As Weis points out, it's important to recognise that such a transfer is, in no sense, any kind of technological backward step.

> On the contrary, to significantly increase the scale of organic and near organic practices will require much more scientific research and training geared towards better understanding how agro-ecosystems operate and how key dynamics can be selectively enhanced. (2007, p. 170)

This highlights the fact that an involvement of significantly greater numbers of people in labour intensive agriculture in no way represents a move away from brain work and into unskilled, unchallenging manual work. As Weis (2007, p. 175) says, the new farming economy involves a 'fusion of physical exertion in the open air with the varied intellectual challenges associated with managing agro-ecosystems and enhancing soils'.

Of course, it is precisely because many developing world economies are effectively controlled by large landholders at home and abroad that they have failed to take any such steps. They have failed to provide a Cuban level of employment, health and welfare provision. In these cases, there will need to be considerable pressure from below – from the landless people themselves and their supporters – at home and abroad to institute the necessary changes.

To the extent that developing countries remain dependent upon agricultural exports, there needs to be a massive extension of something like the current fair trading system to replace neoliberal free trade. As noted in Chapter 5, fair trade certification 'guarantees minimum prices for producer organisations', based on 'local economic conditions and covers production costs plus provision for household members to enjoy a decent living standard, and the cost of farm improvements... and of belonging to a farmers' co-operative' (Litvinoff and Madeley, 2007, pp. 46–7). Indeed, the rapid growth of the fair trade movement involves significant progress for increasing numbers of poorer farmers. However, it does not help the vast numbers of people who have no land upon which to establish such cooperatives.

The issue of lack of industrial production can be addressed by developed world support for a return to import substituting industrial development and for focused development of particular 'cutting edge' industrial manufactures for export. This requires active state intervention in protecting, subsidising, regulating and nurturing appropriate industrial development. There are plenty of items that need to be manufactured, beyond key elements of infrastructure, for such nations themselves, including replacements for all the unsustainable systems currently in operation around the rest of the world (Chang, 2007, p. 213).

The issue of mineral dependency can be addressed by nationalisation of mineral deposits and by a return to the much more effective tax regimes that existed in earlier decades. As Moody points out, this could include

> raising withholding taxes and putting a tax on dividends and upon the mines themselves. Governments might also increase export and import duties, abolish capital allowances, prohibit the offshore retention of earnings and impose 'windfall' taxes during times of larger corporate profit-taking. (Chang, 2007, p. 55)

Most importantly, governments need to 'abolish tax holidays' for mining corporations, and charge royalties based upon 'a deposit's in situ value' rather than based upon profits or gross sales (Chang, 2007, p. 56). The revenues need to be used to drive development in other areas, constructing a sustainable and healthy agricultural and industrial foundation for the future.

Conclusion

Honest consideration of such proposals shows that the obstacles to their implementation are not ethical or technical. Such changes are ethically defensible and technically possible. They are probably also necessary in order to avoid exponential increases in inequality, poverty, misery, famine and conflict. Problems of implementation derive largely from:

1. The ruthless resistance of the major beneficiaries of the current immoral and unsustainable system, including deployment of armed force against any such challenge.
2. The past success of such beneficiaries in confusing and misleading large numbers of people throughout the world – about the issues involved – through their control of major means of communication.
3. The massive inertia of the political and economic system of institutions and ideas already in place.

On the positive side, recent events have seriously shaken both the major institutions and the ideological foundations of the neoliberal consensus. Responses to the financial meltdown and to climate change are still in process. As Mason argues

> the dominance of finance, derivatives and a debt-fuelled growth model is in reality only a little over 10 years old. Likewise, the global imbalances really took off only in the wake of the 1997 crisis. There is, in other words, nothing permanent... about the current shape of the global economy: (Mason, 2009, p. 172)
>
> A socialised banking system plus redistribution is... the ground on which the most radical of the capitalist re-regulators will coalesce with social justice activists. And it may go mainstream if the only alternative is seen to be low growth, decades of debt imposed stagnation, or another re-run of this crisis a few years down the line. It is also possible that a socialised banking system, by allowing the central allocation of financial resources, could be harnessed to the rapid development and large scale deployment of post-carbon technologies. (Mason, 2009, pp. 163–4)

A number of political parties, trade unions and non-government organisations (NGOs) around the world, including some green and socialist parties, have long espoused and campaigned for these sorts of changes. They just have to keep plugging away. The expansion of new channels of communication, via the internet, bypassing established media monopolies, offers hope for future progress.

Most of the reform proposals noted here have the advantage that they can be instituted by degrees and in piecemeal fashion rather than as an all or nothing 'revolution' package. Small steps towards greater democratisation, grassroots empowerment and regulation of corporate operations are better than no steps. On the other hand, the bigger the steps, the bigger, and more brutal, the likely resistance.

In Latin America, in particular, pressures from both political leaders and from mass movements of peasants and workers are pushing in positive directions. At the same time, such movements also demonstrate the intensification of problems and of opposition in response to such initiatives.

We have already seen how Cuba has led the way in a number of important areas. In recent years mass peasant movements, including the Zapatista

National Liberation Army in Mexico, and La Via Campesina, the growing world peasant federation, have engaged in direct action in support of land reform.

Specifically anti-neoliberal governments supposedly actively committed to nationalisation of major productive resources and large-scale land reform have been elected throughout Latin America (Panitch and Leys, 2007, p. 144). While some of these regimes have completely failed to follow through with any such challenge to neoliberal policies and institutions, Chavez in Venezuela (since 1998), Morales in Bolivia (since 2005) and Correa in Ecuador (since 2006) have attempted to create an actively anti-neoliberal regional bloc with significant nationalisation, land reform and empowerment of local communities and grass roots activists.

Bolivia provides a useful case study of possibilities and obstacles. Morales came to power promising to nationalise oil and gas fields and redistribute 20 million hectares of land to 2.5 million landless peasants (out of a total population of 9 million). Although millions of acres of mainly state owned land have been handed over to indigenous and landless communities, the Morales regime has failed to appropriate any significant amount of the greater part of the nation's productive land still held by the handful of big plantation owners – even land they have not cultivated. Indeed, Morales has continued to rely on large-scale agricultural exports and World Bank loans to provide foreign currency to fund imports, and maintain economic and political 'stability'.

Despite this, with support of some police and soldiers, big land owners in the east of the country have organised paramilitary groups to resist such redistribution through murder and intimidation. This has included attacks on camps of peasants who have left the latifundia labour force to squat and work upon other unused land.

All of this goes to highlight the crucial interdependencies of developed and developing economies in a globalised world, and the need for developing world reformists to get away from reliance upon agribusiness exports for necessary foreign currency. It also shows the kind of help that is needed from the developed world in order to assist further stages of reform.

Any supporters of equality of opportunity would presumably reject the current restriction of movement and migration of poor and dispossessed people around the world. A sudden and complete removal of such restriction would still discriminate against those poorer people unable to take advantage of it, and could undermine the opportunities offered in the developed world through sudden and massive pressure on jobs and welfare systems. It is hardly likely to be endorsed by governments under the sway of private wealth. On the other hand, massive population movements, perhaps driven by increasing climate change, could also provide the stimulus required for the sorts of reforms considered here.

Bibliography

Adams, M., 2005, 'FDA Censorship, Suppression of its Own Scientists is Routine, Survey Reveals', *Natural News*, 27 February, http://www.naturalnews.com/005032.html
'"Alien" Genes Escape into Wild Corn', *New Scientist*, (2696), 7, 21 February 2009.
Angell, M., 2005, *The Truth About the Drug Companies*, Scribe, Melbourne.
Archer, D., 2009, *The Long Thaw*, Princeton University Press, Princeton.
Arnst, C., 2009, 'Feds Hit Pfizer with $2.3 Billion Fine', *Business Week*, 2 September, http://www.businessweek.com/bwdaily/dnflash/content/sept2009/db2009092_913433.htm
Ashton, J. and Laura, R., 1998, *The Perils of Progress*, UNSW Press, Sydney.
Australian Government, 2008, Carbon Pollution Reduction Scheme White Paper, 15 December, modified 7 May 2009, http://www.climatechange.gov.au/whitepaper/foreword.html
Barret, D., 2009, 'Pfizer to Pay Record 2.3B Penalty for Drug Promos', Associated Press, 2 September.
Bekoff, M. and Pierce, J., 2009, *Wild Justice*, University of Chicago Press, Chicago.
Bello, W., 2009, *The Food Wars*, Verso, London.
Beresford, Q., Bekle, H., Philips, H. and Mulcock, J., 2004, *The Salinity Crisis*, University of Western Australia Press, Perth.
Born, L., 2004, 'Drugging America to Death', *International Socialist Review*, 33, Jan–Feb.
Borrell, B., 2009, 'A Medical Madoff: Anesthesiologist Faked Data in 21 Studies', *Scientific American*, 10 March, http://www.scientificamerican.com/article
Bremner, M., 1999, *Genetic Engineering and You*, HarperCollins, London.
Burdon, R., 2003, *The Suffering Gene*, Zed Books, London.
Caldicott, H., 2006, *Nuclear Power is Not the Answer*, Melbourne University Press, Melbourne.
Caldicott, H., 2009, *If You Love This Planet*, Norton, New York.
Chang, H.-J., 2007, *Bad Samaritans*, Business Books, London.
Chang, H.-J. and Grabel, I., 2005, *Reclaiming Development*, Zed Books, London.
Chetley, A., 1996, *Problem Drugs*, Stirling Books, Old Noarlunga.
Cooper, R. S., Kennelly, J. and Ordunez-Garcia, P., 2006, 'Health in Cuba', *International Journal of Epidemiology*, 35 (4), 817–24, http://ije.oxfordjournals.org/cgi/content/full/35/4/817

Dahlgren, G. and Whitehead, M., 1991, *Policies and Strategies to Promote Equity in Health*, Institute for Future Studies, Stockholm.
Davidson, P., 2009, *John Maynard Keynes*, Palgrave Macmillan, Houndmills.
De Feyter, K., 2005, *Human Rights*, Zed Books, London.
De Waal, A., 2006, *Aids and Power*, Zed Books, London.
Diesendorf, M., 2007, *Greenhouse Solutions with Sustainable Energy*, UNSW Press, Sydney.
Diesendorf, M., 2009, *Climate Action*, UNSW Press, Sydney.
Dirnfeld, V., 1996, 'The Benefits of Privatisation', *Canadian Medical Association Journal*, 155 (4), 407–10.
Eatwell, J. and Taylor, L., 2000, *Global Finance at Risk*, Polity Press, Cambridge.
Flannery, T., 2009, *Now or Never*, Black Inc, Melbourne.
Flynn, M., 2003, *Human Rights in Australia*, Butterworths, Sydney.
Foley, D., 2006, *Adam's Fallacy*, Harvard University Press, Cambridge, Mass.
Funnell, W., Jupe, R. and Andrew, J., 2009, *In Government We Trust*, UNSW Press, Sydney.
George, S., 2004, *Another World is Possible, If . . .* , Verso, London.
Gigerenzer, G., 2002, *Reckoning with Risk*, Allen Lane, London.
Gostin, L. O., Areen, J., King, P. A., Goldberg, S. and Jacobson, P. D., 2005, *Law, Science and Medicine*, Foundation Press, New York.
Graham, H., 2007, *Unequal Lives*, Open University Press, Maidenhead.
Greene, M. F., 2006, *There Is No Me Without You*, Bloomsbury, London.
Greenpeace Australia, 2009, *Truefood: The Guide*, http://www.truefood.org.au/truefoodguide
Grey, B., 2009, 'Wall Street Celebrates Government Windfall for Banks and Big Investors', *World Socialist*, 24 March, http://www.wsws.org/articles/2009/mar2009/toxi-m24.shtml
Hahnel, R., 2002, *The ABCs of Political Economy*, Pluto Press, London.
Hansen, J., 2009, *Storms of My Grandchildren*, Bloomsbury, London.
Harman, C., 1984, *Explaining the Crisis*, Bookmarks, London.
Harremoes, P., Gee, D., MacGarvin, M., Stirling, A., Keys, J., Wynne, B. and Guedes Vaz, S. (eds), 2002, *The Precautionary Principle*, Earthscan, London.
Head, M. and Mann, S., 2009, *Law in Perspective*, 2nd edn, UNSW Press, Sydney.
Ho, M.-W., 1999, *Genetic Engineering; Dream or Nightmare?* Gateway, Dublin.
Horton, R., 2004, 'Vioxx, the Implosion of Merck and Aftershocks at the FDA', *The Lancet*, Published online 5 November, http://image.thelancet.com/extras/04cmt396web.pdf
Irvine, B., 2002, Background Briefing on US Healthcare Based on G.-M. Turner, *Health Policy Consensus Group Report*, http://www.civitas.org.uk/pdf/USABrief.pdf
Keen, S., 2001, *Debunking Economics*, Pluto Press, Sydney.
Kelly, M., 2003, *The Divine Right of Capital*, Berrett-Koehler, San Francisco.
Kempf, H., 2009, *How the Rich are Destroying the Earth*, Finch Publishing, Sydney.
Kevin, T., 2009, *Crunch Time*, Scribe, Melbourne.
Kohonen, M. and Mestrum, F. (eds), 2009, *Tax Justice*, Pluto Press, London.
Korczak, V., 2005, 'TGA Transparency and Independence', *Choice*, http://heritage.choice.com.au/files/f123731.pdf
Krimsky, S., 2004, *Science in the Private Interest*, Rowman and Littleman, Lanham.

Krugman, P. and Wells, R., 2006, 'The Health Care Crisis and What To Do About It', *New York Review of Books*, 53 (5), 23 March; http://www.nybooks.com/articles/18802

Kuhse, H. and Singer, P., 2006, *Bioethics; An Anthology*, 2nd edn, Blackwell, Oxford.

Kunst, A. E., Bos, V., Lahelma, E. et al., 2005, 'Trends in Socio-economic Inequalities in Self-assessed Health in Ten European Countries', *International Journal of Epidemiology*, 34, 295–305.

Lannin, S., 2009, 'Pfizer Fine Should Act as Deterrent', *The World Today*, 3 September, http://www.abc.au/news/stories/2009/09/03/2636083.htm

Lazerou, J., Pomeranz, B. and Corey, P., 1998, 'Incidence of Adverse Drug Reactions in Hospitalized Patients; a Meta-analysis of Prospective Studies', *Journal of the American Medical Association*, 279, 1200–5.

Levine, A., 2002, *Engaging Political Philosophy*, Blackwell, Oxford.

Litvinoff, M. and Madeley, J., 2007, *50 Reasons to Buy Fair Trade*, Pluto Press, London.

Lohmann, L., 2006, 'Carbon Trading; Solution or Obstacle?', http://www.thecornerhouse.org.uk/summary.shtml?x=561553

Lynas, M., 2008, *Six Degrees*, Harper Perennial, London.

Mappes, T. and DeGrazia, D., 2006, *Biomedical Ethics*, 6th edn, McGraw Hill, Boston.

Marmot, M., 2004, *Status Syndrome*, Bloomsbury, London.

Maslin, M., 2009, *Global Warming; A Very Short Introduction*, Oxford University Press, Oxford.

Mason, P., 2009, *Meltdown*, Verso, London.

Mehrotra, S. and Delamonica, E., 2007, *Eliminating Human Poverty*, Zed Books, London.

Miliband, R., 1991, *Divided Societies*, Oxford University Press, Melbourne.

Miliband, R., 1994, *Socialism for A Sceptical Age*, Polity Press, Cambridge.

Moggridge, D. (ed.), 1982, *The Collected Writings of John Maynard Keynes*, Volume 21, Macmillan, London.

Moody, R., 2007, *Rocks and Hard Places*, Zed Books, London.

Moore, M., 2007, *Sicko*, The Weinstein Company, Dog Eat Dog Films.

Morris, C., 2009, *The Two Trillion Dollar Meltdown*, Black Inc, Melbourne.

Morton, A., 2009, 'Greenhouse pledges way too low: UN', *The Age*, 13 June.

Moynihan, R., 1998, *Too Much Medicine*, ABC Books, Sydney.

Moynihan, R. and Cassels, A., 2005, *Selling Sickness*, Allen and Unwin, Sydney.

National Coalition of Healthcare, Health Insurance Coverage, www.nchc.org/facts/coverage.st=htmal

Nickel, J. W., 2007, *Making Sense of Human Rights*, 2nd edn, Blackwell, Malden.

Noonan, D. and Nazario, B., 2005, 'Holes in US Drug Safety Net', http://www.skeptically.org/quackery/id26.html

Norman, R., 1983, *The Moral Philosophers*, Clarendon Press, Oxford.

Norman, R., 1987, *Free and Equal*, Oxford University Press, Oxford.

Norman, R., 1998, *The Moral Philosophers*, 2nd edn, Oxford University Press, Oxford.

Nossal, G. J. V. and Coppel, R. L., 2002, *Reshaping Life*, 3rd edn, Melbourne University Press, Melbourne.

Panitch, L. and Leys, C., 2007, *Global Flashpoints*, Merlin Press, London.

Pannell, D., Lefroy, T. and McFarlane, D., 2000, 'A framework for prioritising government investment in salinity in Western Australia', *Salinity Policy Plan* (unpublished).

Parker, C. and Evans, A., 2007, *Inside Lawyers' Ethics*, Cambridge University Press, Melbourne.

Pearce, F., 2006, *When the Rivers Run Dry*, Transworld Publishers, London.

Piñeyro-Nelson, A., Van Heerwaarden, J., Perales, H. R., Serratos-Hernández, J., Rangel, A., Hufford, M. B., Gepts, P., Garay-Arroyo, A., Rivera-Bustamante, R. and Álvarez-Buylla, E. R., 2009, 'Transgenes in Mexican Maize: Molecular Evidence and Methodological Considerations for GMO Detection in Landrace Populations', *Molecular Ecology*, 18 (4), 750–61, February.

Pressman, S., 1999, *Fifty Major Economists*, Routledge, London.

Rawls, J., 1972, *A Theory of Justice*, Harvard University Press, Harvard.

Resnik, D. B., 1998, *The Ethics of Science*, Routledge, London.

Roberts, D., 2008, *Human Insecurity*, Zed Books, London.

Roberts, P., 2008, *The End of Food*, Bloomsbury Publishing, London.

Robertson, G., 2009, *The Statute of Liberty*, Vintage Books, Sydney.

Robinson, T. R., 2005, *Genetics for Dummies*, Wiley, Hoboken.

Rosset, P. M., 2006, *Food is Different*, Zed Books, London.

Rouvroy, A., 2008, *Human Genes and Neoliberal Governance*, RoutledgeCavendish, Abingdon.

Rubin, I., 1979, *A History of Economic Thought*, InkLinks, London.

Rudd, K., 2009, 'The Global Financial Crisis', *The Monthly*, 12, http://www.themonthly.com.au/node/1421

Schapiro, M., 2007, *Exposed*, Chelsea-Green Publishing, White River Junction.

Self, P., 2000, *Rolling Back The Market*, Macmillan Press, Basingstoke.

Shah, S., 2006, *The Body Hunters*, The New Press, New York.

Shiva, V., 1991, 'The Green Revolution in the Punjab', *The Ecologist*, 21 (2), March–April, http://livingheritage.org/green-revolution.htm

Shiva, V., 2005, *Earth Democracy*, South End Press, Cambridge, MA.

Shiva, V., 2008, *Soil Not Oil*, Spinifex Press, Melbourne.

Short, S. and McConnell, C., 2000, *Extent and Impacts of Dryland Salinity*, Agriculture WA, Perth.

Shutt, H., 2001, *A New Democracy*, Zed Books, London.

Singer, P. and Mason, J., 2006, *The Ethics of What We Eat*, Text Publishing, Melbourne.

Singh, K., 2005, *Questioning Globalization*, Zed Books, London.

Sjolander, A.-C., 2005, *The Water Business*, Zed Books, London.

Smith, J. M., 2007, *Genetic Roulette*, Gene-Ethics, Melbourne.

Soros, G., 2008, *The New Paradigm for Financial Markets*, PublicAffairs, New York.

Spake, A., 2004, 'A Sick Agency in Need of a Cure?' *USN*, http://health.usnews.com/usnews/health/articles/041213/13fda.htm

Spencer, N., 2008, 'Food Prices Continue to Rise Worldwide', *World Socialist*, February, http://www.wsws.org/articles/2008/feb2008/food-f25.shtml

Spratt, D. and Sutton, P., 2008, *Climate Code Red*, Scribe, Melbourne.

Steckx, S., 2005, 'Can Drug Patents be Morally Justified?' *Science and Engineering Ethics*, 81, 89.

Stiglitz, J., 2006, *Making Globalization Work*, Allen Lane, Camberwell.

Stiglitz, J. and Bilmes, L., 2008, *The Three Trillion Dollar War*, Allen Lane, Camberwell.

Stilwell, F., 2002, *Political Economy*, Oxford University Press, Melbourne.
Stillwell, F. and Jordan, K., 2007 *Who Gets What? Analysing Economic Inequality in Australia*, Cambridge University Press, Melbourne.
Tickell, O., 2008, *Kyoto 2*, Zed Books, London.
Turner, G., 2008, *The Credit Crunch*, Pluto Press, London.
Uretsky, S., 2008, 'HealthCare in the United States, MedHunters', http://www.medhunters.com/Articles/healthcareInTheUsa.html *and* http://www.cdc.gov/cancer/NBCCEDP/about.htm
Usdin, S., 2007, *The No-Nonsense Guide to World Health*, New Internationalist Publications, Oxford.
Walker, S., 2007, *Biotechnology Demystified*, McGraw-Hill, New York.
Waluchow, W. J., 2003, *The Dimensions of Ethics*, Broadview Press, Peterborough.
Weis, T., 2007, *The Global Food Economy*, Zed Books, London.
Weiss, L., Thurbon, E. and Matthews, J., 2004, *How to Kill a Country*, Allen and Unwin, Sydney.
Wilkinson, R. G., 2005, *The Impact of Inequality*, Routledge, London.
Wilkinson, R. and Pickett, K., 2009, *The Spirit Level*, Allen Lane, London.
Wilson, J., 2008, 'Wall Street Grain Hoarding Brings Farmers, Consumers Near Ruin', *Bloomberg*, 28 April, http://www.bloomberg.com.au/apps/news?pid=20601087&sid=aDZej7GJjpjM&refer=home
Windisch, M., 2009, 'Rudd's Carbon Trading, Locking in Disaster', *Green Left Weekly*, 23 May, http://www.greenleft.org.au/2009/796/41003

Index

abortion 19–20, 98, 222–3
ACCC *see* Australian Competition and Consumer Commission
accelerator driven thorium reactors 196
addiction (drug) 224–6
adverse selection 200, 203
advertising 116
   drug promotion 254
   of junk foods 119
   and obesity 140–1
agriculture
   agricultural biotechnology 160
   agricultural exports 266
   antibiotics use 165–6, 226
   biodiverse farms 126
   capitalist farming 125, 135
   cash cropping 136
   demand for agricultural commodities 120
   dryland salinity 128–9, 130, 162
   and excess nitrogen 130–1, 133
   industrial agriculture 121–3, 266
   irrigation 130–2, 144
   labour saving technologies in farming 118
   native vegetation clearance 129
   organic farming 139–40
   soil's carbon storage 134
   and subsidies 137–8, 180
   subsistence farming 122, 135–7
   sustainable farming techniques 125–6
   swidden agriculture 124, 126
   see also food production
agrochemicals 140
aid 262–4
AIDS 229, 241–2, 244
ALLHAT (clinical trial) 249

animals 16
   animal rights 16, 143–4, 237
   animal testing – drugs 236–8
   CAFOs and 'processing' of animals 122
   the Enviropig 166
   factory farmed animals 126–7, 143–4
   GM/GE animals 125, 164–6
   and growth hormone genes 164, 165–6
   and human pharmaceuticals production 148
   moral rights 237
   transgenes 147–8
   use of antibiotics in farm animals 119
antibiotics 165–6, 217–18, 221–2, 226
   antibiotic resistance 119, 126–7, 154, 156, 217–18
   overuse 226
   use in intensive industrial farming 226
appropriation 27–8
artificial NPK fertilisers 127–8
asbestosis 111–12
assets 53, 87–8
Australian Competition and Consumer Commission (ACCC) 52
Australian Securities and Investments Commission (ASIC) 52
   consumer protection role 52
   efficient markets hypothesis 52, 80
autonomy rights 12–13

barbiturates 224–5
belief systems 20–2
benzodiazepines 225
Bextra 254
bias 246–7
biodiverse farms 126

274

bio-fuels 189–92
  bagasse 190
  ethanol and biodiesel 190
borrowing/lending 84–7, 88
British National Health Service (NHS) 107, 209
brownfield investment 49

cadmium 142
CAFOs *see* concentrated animal feeding operations
CaMV promoter 156
cap and trade (carbon trading) 179–86, 187
capital
  capital account liberalisation 50
  'effort and sacrifice' concept 90–2
  foreign direct investment 49–50, 55
  international capital flows 49–50
capitalism 12, 93
  capitalist farming 125, 135
  capitalist market system 25, 29, 30
  food production and capitalist competition 118–20
  see also liberalism
carbon credits 184, 186
carbon leakage 185
carbon offsets 181–2
carbon storage 134
carbon trading 178–86, 187
carcinogenic organic compounds 173
cash cropping 136
cDNA *see* complementary DNA
CDO *see* collateralised debt obligations
central banking 46–7, 67–8, 83, 86–7
  China's reserves of safe currencies 76
  focus on inflation 47
Chernobyl meltdown 195
chronic disease 108–9
civil law 29
class inequalities 22–3, 113
class structure 109–13
clean coal 192–7
climate change 139, 174, 185, 186
  adaptability to 176
  and reduction of greenhouse gas emissions 172
  and sustainability 133–4
clinical trials (drugs) 238–9, 249
clones 154
CO_2 emission 171
  and biomass 189–90
  capture 192

  future CO_2 emissions 174
  and nuclear energy 194
  pyrolysis 191–2
  rainforest absorption of CO_2 177
  sequestration 192–3
  storage safety 193
coal mining 172–3, 192–7
collateralised debt obligations (CDO) 86
commodities 120–1
  demand for agricultural commodities 120
  health care as a commodity 198
  hedge fund withdrawal 121
Commonwealth Scientific and Industrial Research Organisation (CSIRO) 166, 172
communications 23, 262
comparative advantage theory 47–9, 72–3, 119
competition 37, 39, 42, 65
  competitive markets 37, 64–5
  and developing nations 70–1
  food production and capitalist competition 118–20
  free market competition 61–2
  and innovation 64
  neoclassical theory – perfect competition 40–1
  problems 60–95
  state control of China's resources 75–7
complementary DNA (cDNA) 168
concentrated animal feeding operations (CAFOs) 122
conflict of interest 246–7, 249–50, 253
consumer protection 52
consumption/consumers
  consumer choice and health care 200
  consumer nutrition awareness 141
  consumption patterns 115–16
  food overconsumption 141
  meat consumption 122, 126–7
  medical services overconsumption 215–16
  water consumption 139
contract law 12, 13
contract research organisations (CROs) 238
convertibility (of currency) 53–4, 74–5, 76
cooperation 33–4
cooperatives 259
copyrights *see* intellectual property rights
corporations
  control of education and research 63
  control of patents 127
  and private ownership of resources 40–2

Council of Europe's Convention on Human Rights and Biomedicine 203
credit 88
CROs *see* contract research organisations
CSIRO *see* Commonwealth Scientific and Industrial Research Organisation
Cultural Revolution 138
currency
   convertibility of 53–4, 74–5, 76
   currency boards 54
   demand for 53
   devaluation 74, 75
   and exchange rates 74–5
   reserve currency 77
   safe currencies 76

debt 85–7, 264
Declaration of Helsinki 240–1, 243
deficits
   deficit spending – 'stimulation' 88
   trade deficits 77
deforestation 126, 133, 172, 174, 184
demand 38–9, 40, 64
   and ACCC 52
   for agricultural commodities 120
   for currency 53
   and food production 118, 121
   Friedman's theories 44–5, 46, 199
   Keynes' theories 44–6
   price elasticities of demand 74
democracy 24, 62–4, 71
   democratic rights 12
   democratisation 259, 261–2, 263
   limitations in developing world 64
   and politics 261–2
   see also liberalism
deontological ethical theories 11–12, 14, 16
deregulation 75, 134–5
   financial deregulation 50–2, 78–80, 82–3
   of investment banking 120–1
   versus regulated systems 52
derivatives 51, 83–4
   global derivatives market 84
devaluation (of currency) 74, 75
diabetes 102, 117, 249–50
diethylstilbestrol (DES) 222–4
   carcinogenic and teratogenic effects 223–4
   role in spontaneous abortions 222–3
difference principle 33
distributive justice 28, 29

'contractarian' theory of distributive justice 31–4
dividend imputation credit system 69
division of labour 37
DNA 149–50, 152, 154–60, 164, 167
   complementary DNA (cDNA) 168
Doha Declaration 229–30, 235
downturns 61
drugs
   AIDS 229, 241–2, 244
   animal testing 236–8
   antibiotics 119, 165–6, 217–18, 221–2, 226
   barbiturates 224–5
   benzodiazepines 225
   Bextra 254
   clinical trials 238–9, 249
   development 235–6
   diethylstilbestrol 222–4
   drug companies and free trade agreement 233–5
   drug promotion 254
   drug testing 245–8
   essential drug access 229–30
   FDA approval 245
   FDA issues 251–2
   financial conflict of interest and bias 246–7, 249–50
   flecainide 247–8
   growth of private pharmacies 217
   HIV vaccine 242
   human research ethics 239–41
   human testing 238–9
   India's regulation of drugs/medicine 218
   informed consent 250–1
   Lipitor 254
   'litigation strategy' 248
   marketing 231–2
   'me too' drugs 73, 245, 251
   monopoly rights for brand name drugs 228
   non-nucleotide reverse transcriptase inhibitors 242
   off-label prescribing 254
   Paxil (Aropax) 225
   penicillin 222
   prevastin sodium 232–3
   prices 215, 234–5, 244–5
   privatisation and intellectual property 227–30
   problem drugs 221–2
   protease inhibitors 242

psychotrophics and addiction 224–6
publicly funded research 248–50
R&D budgets 244
Retrovir (AZT) 241–2, 243
Rezulin 249–50
scientific evidence – risk reduction 232–3
selling sickness 230–2
sulfanilamide 221–2
surrogate end points 247
thalidomide 227
Viagra 230, 254
waivers 252–3
dryland salinity 128–9, 130, 162
see also irrigation
duty concept 11–13

ecology 124–7
effects of global warming 171–2
environmental impact of industrial agriculture 125
see also climate change
economies/diseconomies of scale 64, 65, 66, 257–8
education 19, 63, 112, 216
efficiency concept 41
efficient markets hypothesis 52, 80
'effort and sacrifice' concept 90–2
egalitarianism 103–4
El Niño 175
Emission Trading System (ETS) 182
empathy 10–11
employment 114–15
see also unemployment
energy
Australia's greenhouse gas emissions 172, 184–7
bioenergy 189–92
clean coal 192–7
energy of coal – coal-fired power stations 173
energy production from fossil fuels 135, 170
energy saving, wind and solar power 187–9
geothermal power 187
and the greenhouse effect 170–97
hydroenergy projects 188
nuclear energy 193–7
oil and gas depletion 171, 173–4
use in manufacturing and construction 170
wave power technologies 188

Environmental Protection Agency (EPA) 158
Enviropig 166
equality of opportunity 17, 30, 32, 92–3
equilibrium price 37, 39, 53
ethanol 190
ethics 10–11, 239–41
and animal testing 236–8
deontological theories 11–12, 14, 16
ethical issue and empirical/factual issue relationship 19–20
GM – ethical issues 163
issues 19–20, 163
problems of just deserts 88–90
responses to issues of ethical justification of free markets 56–8
social, personal and professional ethics 18–19
teleological theories 11
Ethics in Government Act 253
ETS *see* Emission Trading System
euro, the 77
exchange rates 53–4, 74–5
export 74, 76, 77–8, 263
agricultural exports 266
capitalist (export) farming 125, 135
and plantation production 121
uranium exports 193
US exports of surplus grain 134–5

factory farmed animals 126–7, 143–4
fair trade movement 266
farming *see* agriculture
FDA *see* Food and Drug Administration
FDI *see* foreign direct investment
Federal Reserve 88
female abortion 98
'fight and flight' response 115
financial deregulation 50–2, 78–80, 82–3
financial liberalisation 50–2, 79
fiscal policies 43–5
flecainide 247–8
floating exchange rates 53–4
flooding 175
Food and Drug Act 227
Food and Drug Administration (FDA) 142–3, 238–9, 245, 248, 250
drug promotion 254
FDA issues – drugs 221, 224, 251–2
GM issues 160–1
waivers 252–3
food-borne infections/pathogens 117
antibiotic resistant strains 119, 126–7

Food, Drug and Cosmetic Act (1938) 222
food production 145
   animal rights issues 143-4
   and capitalist competition 118-20
   China 138-9
   concentration and centralisation of farming activities 118
   and corporate 'free' markets 117
   corporation control 127
   crop yield and GM 146-7, 162
   Cuba 139-40
   death of the peasantry 135-7
   and demand 118, 121
   developments 118-21
   food health issues 142-3
   'food insecurity' 117, 176
   food-related protests and riots 121
   free trade debate 137-8
   the global South 123-4
   grain hoarding 121
   the Green Revolution 98, 124, 125, 126
   high-yield plant strains and Haber-Bosch (fertiliser) system 122
   homogenisation 127-8
   industrial agriculture 121-3, 266
   labelling of GM foods 159-60
   land consolidation and subsistence farming displacement 122
   malnutrition, hunger, and production excess 119
   obesity issues 140-2
   overproduction 122-3, 134
   plantation production 121
   'processing' of animals 122
   role of IMF and World Bank 134-5
   safety of GM foods 160-2
   speculation about agricultural commodities 120-1
   sustainable farming techniques 125-6
   toxicity of GM foods 158-9
   vulnerability of poorer populations to food price rises 135
   and water 144-5
   water pollution and dense livestock 131
food safety 148, 160-2
Food Standards Australia New Zealand (FSANZ) 148, 161-2
foreign direct investment (FDI) 49-50, 55
fossil fuels 125, 126, 133
   energy production from fossil fuels 135, 170
   future usage 174
   and greenhouse gas emissions 171
   see also carbon trading
free currency markets 53-4
free international trade 47-8, 71-2, 135, 233-5
   free trade debate in food production 137-8
   specialisation 71-2
free markets 25-7, 56-7, 149
   consumer choice and health care 200
   ethical justification of 56-8
   expansion 136
   free internal market – China 138
   free market competition 61-2
Free Trade Agreement (FTA) 233-5
Friedman's theories 44-5, 46, 199
FSANZ *see* Food Standards Australia New Zealand
fusion power 197

Garnaut Climate Change Review 185, 186
gas 171
   see also CO_2 emission; greenhouse effect; oil
GDP *see* gross domestic product
General Agreement on Tariffs and Trade (GATT) 55
genes 149-51, 154-60
   gene flow 157-8
   growth hormone genes 164, 165-6
   Human Genome Project 150-1
   insertion into animal genomes 164
genetic modification (GM)
   antibiotic resistance 126-7, 154, 156, 217-18
   antibiotics levels in dairy products 165-6
   bans and restrictions 152-3
   Bt toxin 155-6, 158, 159
   CaMV promoter 156
   clones 154
   contamination/pollution 153, 157
   DNA 149-50, 152, 154-60, 164, 167
   ethical issues 163
   FDA issues 160-1
   feeding the hungry 162-3
   food safety issues 148, 160-2
   the future 169
   GM crops 125, 126-7, 151-60, 162-3
   GM/GE animals 125, 164, 166
   GM traits 152
   herbicide tolerant (HT) crops 152, 156

Human Genome Project 150-1
insect resistance 152
L-tryptophan supplement 159
labelling of GM foods 159-60
life patents 166-9
NAG compound 156
the promise 146-9
safety of GM foods 160-2
StarLink Bt toxin 159
toxicity of GM foods 158-9
transgenes 147-8, 154-60, 164
genetic predispositions 150
genetic property rights 166-9
genetic testing 203, 204
genital mutilation 104-5
geothermal power 187
Glass–Steagall Act 82
global life expectancy *see* life expectancy
global warming 117, 125, 133, 139, 171
   5 degrees of global warming 177-8
   'belt of habitability' 177
   and bio-fuels 189-92
   'the cull' 178
   disease transmission 176
   effects 171-2
   hurricanes and El Niño 175
   melting of ice 175
   nuclear energy 193-7
   rainfall and flooding 175
   rainforest absorption of CO_2 177
   sudden increases in warming 176
GM *see* genetic modification
GNI *see* gross national income
GNP *see* gross national product
government
   central government and land reform 265
   deficit spending – 'stimulation' 88
   Ethics in Government Act 253
   fiscal policies 43-5
   government intervention (monopolies) 39-40, 42-3
   greenhouse gas reduction targets and infrastructure 186-7
   health care safety net 200
grain hoarding 121
Green Revolution (GR) 98, 124, 125, 126
   GR in the Punjab 132-3
   GR rice 130
   and growth rate of total crop production 132

interpretation 148
and irrigation 131
greenfield investment 49, 75
greenhouse effect
   Australia's greenhouse gas emissions 172, 184-7
   components 172
   energy saving, wind and solar power 187-9
   global warming and fossil fuels 171
   greenhouse gas emissions 133-4, 171, 172
   infrastructure 186-7
   Kyoto Protocol 178-84
   sea level rise predictions 174-5
gross domestic product (GDP) 29, 56, 99-100, 125, 208, 263
gross national income (GNI) 97, 103
gross national product (GNP) 103, 219
growth hormone genes 164, 165-6

Haber-Bosch (fertiliser) system 122
Hayek, Friedrich 39-40, 58
health care 12, 13
   British National Health Service 107, 209
   in Canada 208-9, 210
   cost of 217
   food health issues 142-3
   genetic predispositions 150
   in Germany 209
   government safety net 200
   health care as a commodity 198
   health risks/dangers of coal mining 172-3
   human health and meat consumption 122, 126-7
   inequality and adverse health outcomes 78
   life expectancy 99-100, 104, 106, 213
   loss of access to fresh water 175-6
   market forces and incompetence 199
   moral hazard and medical insurance 199-200
   obesity 109, 114, 117, 140-1, 142, 213
   obstruction of effective health care by patents 169
   physician self-regulation 199
   public funding for 198-9, 200
   public health systems 107, 202, 214-15, 218-20
   in the US 204-6, 213
   World Health Organization health indicators 14

health care provision 198–220
  analysis 214–16
  assessment of private health care (developed world) 212–14
  comprehensive insurance testing – actuarial justice 203–4
  the developing world 210–12
  fee-for-service medicine 206, 211
  free choice 201–3
  health spending 212
  history 206–8
  lifestyle choices and health status 202–3
  managed care system 206–7, 208
  medical services overconsumption 215–16
  Medicare program 205, 214
  neoliberal argument for a complementary private system 210
  neoliberal critique 209–10
  OECD figures 204–5
  OECD systems 208–9
  organisational forms 211
  private provision 198–201, 204–6, 210, 212–14, 216–17, 218
  and profit driven market relations 201–2
  public provision 107, 202, 214–15, 218–20
  'rationing' of public health care 202
  selling sickness 230–2
  transnationalisation of health sector (Latin America) 211–12
health inequalities 96–116
  between countries 96–8
  data on child health 96
  the developed world 113–16
  the developing world 110–13
  epidemiological transition 107–9
  history 98–100
  life expectancy 99–100
  lifestyle factors 113
  and the mining industry 111–12
  more unequal societies are more unhealthy 102–4
  nutrition/health relationship 110
  obesity 109, 114, 117, 140–1, 142, 213
  over time 106–7
  role of class structure 109–13
  'sacrifice areas' 112
  safe water and sanitation 110–11
  and stress 114–16
  within countries 101–2
  women 104–6

health insurance 199–200, 203–4
  actuarial justice principle 203–4
  and adverse selection 200, 203
  employer sponsored insurance (US) 205–6
  and risk 200
  'sickness funds' 209
  'threshold' system 204
Health Maintenance Organization (HMO) 207
hedge funds 83–4, 85–7, 121
herbicides
  herbicide resistance 157
  herbicide tolerant (HT) crops 152, 156
HFC-23 (greenhouse gas) 183
HIV vaccine 242
HSC-23 (greenhouse gas) 183
Human Genome Project 150–1
human research ethics 239–41
human rights 12–13, 16–18, 71, 203
  conflict regarding essential drug access 229–30
  enactment of law 13
  and health care 12
hunger 119, 149
  and GM crops 162–3
  supply shortages and food prices 120
hurricanes 175
hydroenergy projects 188

ideologies 20–2
  and class inequalities 22–3
  ideological function (of society) 20
  and materialism 22
  see also belief systems
IGF-1 see insulin-like growth factor
IMF see International Monetary Fund
import 74, 75, 119
  food import dependence 136
  imported foodstuffs 139
incentives (positive) 94–5
income inequality 102–3
industrial agriculture 121–3, 266
  agricultural price supports and production limits 122–3
  CAFOs and 'processing' of animals 122
  dependence upon irrigation 130
  ecological considerations 124–7
  environmental impact 125
  land consolidation and subsistence farming displacement 122
  plantation production 121

and sustainability 133–4
world market in agricultural produce 121
industrialisation 76, 123, 128
inequalities 96–116, 149
   and adverse health outcomes 78
   class inequalities 22–3, 113
   extent of income inequality 102–3
   and health care provision 216
   see also class structure
infant mortality 97–8, 103
infanticide 98
infectious diseases 108–9, 218
inflation 45–6, 47, 67–8
informed consent 250–1
   informed choice of participation 250–1
infrastructure 70, 264
inheritance 92–3, 261
innovation 64, 73–4
insecticides 127, 132, 158
insulin-like growth factor (IGF-1) 165
intellectual property rights (IPRs) 43, 167, 168
   China's failure to protect 76
   conflict regarding essential drug access 229–30
   drug patents 228–9
   drugs – privatisation and intellectual property 227–30
   industrial designs 228
   TRIPS Agreement 43, 228–9, 235
   see also patents
interest rates 67–8, 77–8, 87
   easy lending 84–5
   low interest rates 76–7, 84
Intergovernmental Panel on Climate Change (IPCC) 174
International Covenant on Social, Economic and Cultural Rights (1966) 13
International Finance Corporation 212
International Monetary Fund (IMF) 42, 50, 54–5, 70, 71
   and currency convertibility 75
   and devaluation 75
   encouragement of transnationalisation 212
   policies 100
   role in food production 134–5
   Structural Adjustment Programs 55
   see also monetary policy; currency
international trade 47–8, 71–2, 74–5
investment 53–4, 72
   in commodities 120–1

foreign direct investment 49–50, 55
greenfield and brownfield investment 49, 75
investment banks 83–4
investment funding 92
and risk 51, 58
invisible hand (of markets) 37–8
IPCC see Intergovernmental Panel on Climate Change
IPRs see intellectual property rights
irrigation 130, 131, 132, 144
   low cost irrigation projects 131
   see also dryland salinity
ischemic heart disease 101–2, 107, 108, 117

job security 115
junk foods 119
just deserts principle 16–18, 28, 29, 57–8, 88–90
justice 27–8, 40
   actuarial justice principle 203–4
   as cooperation 33–4
   law and justice 28–9
   Nozick's distributive justice theories 28
   Rawls' theory of justice 31–4
   retributive justice 29

Kant's theories 11
Keynes' theories 44–6, 72, 80
Kyoto Protocol 178–84

labour 111
   division of labour 37
   food production in China 138–9
   labour costs 77
   labour-saving technologies in farming 118
   rewarding labour as sacrifice 90–1
land
   'belt of habitability' 177
   land consolidation 122
   land ownership 123
   land reform 123–4, 265, 266
law
   civil law 29
   contract law 12, 13
   and human rights 13
   and justice 28–9
   and liberalism 24
   see also legislation
left-wing (social) liberalism 29–31
legal rights 12

legislation
    and domination of the political process 63–4
    drug patents and research 227–8
    legal rights creation 12
    Work Choices legislation 69–70
    see also law
lending/borrowing 84–7, 88
leveraging 83–4
liberalisation 51, 55, 134–5
    capital account liberalisation 50
    financial liberalisation 50–2, 79
    liberalisation of international capital flows 49
liberalism 23–7
    left-wing liberalism 29–31
    right-wing liberalism 25–7
    see also neoliberalism
liberty 27–8, 32, 56
life expectancy 99–100, 104, 106, 213
    GDP and life expectancy relationship 99–100
life patents 166–9
Lipitor 254
liquidity 51, 84
Locke's theories 27–8, 56

mallee planting 191
malnutrition 119, 120, 138
Mandatory Renewable Energy Target (MRET) 186, 187
market prices 89
market relations 24
    capitalist market relations and right-wing liberalism 25, 29
    and health care provision 201–2
    social liberals' view 30, 31
marketing 116, 231–2
Marshall, Alfred 38–9
material function (of society) 20
materialism 22
maternal mortality 104
maximisation principle 14–15
    profit maximisation 41
'me too' drugs 73, 245, 251
mechanisation 37, 39
Medicare program 205, 214
'Medicines Working Group' 234
megabanks 83
minerals 266
mining industry 111–12
    coal mining 172–3, 192–7
    uranium mining 195–6

Ministry of International Trade and Industry (MITI) 66
Minsky, Hyman 80–1
monetary policy 44–5, 46–7, 67–8
money creation 55, 88
monocultures 125, 127–8
    monocultures associated with GM crops 158
    vulnerability of 128
monopolies 38, 39–40, 42–3, 61–2, 258
    democratisation 259, 261–2, 263
    monopoly rights for brand name drugs 228
    private versus public monopoly 65–6
    private wealth and democracy 62–3
    and privatisation 42–3
    see also oligopolies
morality 10–11
    and action 11–12
    and conflict 19
    moral hazard and medical insurance 199–200
    moral rights 237
    social, personal and professional ethics 18–19
    social relationships 13
mortality 97–8
    death rates and egalitarian states 103–4
    and employment hierarchy studies 114–15
    and epidemiological transition 107–9
    infant mortality 97–8, 103
    maternal mortality 104
    'threshold effect' 101
MRET *see* Mandatory Renewable Energy Target

NAFTA *see* North American Free Trade Agreement
NAG compound 156
National Health and Medical Research Council (NHMRC) 236
National Health Service (NHS) 107, 209
National Land and Water Resources Audit 129
National Research Council (NRC) 227
natural price 37–8
natural rate (of payment) 37
neoliberalism 40
    drugs – privatisation and intellectual property 227–30
    and financial deregulation 50–2, 78–80, 82–3
    and floating exchange rates 53–4

and free trade 47–8
and international capital flows 49–50
and IPR protection 43
justification of managed care system 207
and monetary policy 46–7
neoclassical theory of public finance 200–1
neoliberal argument for a complementary private system 210
neoliberal critique of health care provision 209–10
neoliberal economic policies 60–1, 68–70, 141
perfect competition 40–1
and private provision of health care 204–6
and problems of privatisation 66–7
and public health care provision 214–15
push for corporate 'free' markets in foodstuffs 119
and 'rationing' of public health care 202
responses to issues of ethical justification of free markets 56–8
support for user-pays services 198–201
support of regulation 52–3
view of excess medical services consumption 215–16
view of taxation 46
views of risk 51
NHMRC *see* National Health and Medical Research Council
NHS *see* British National Health Service; National Health Service
nitrate fertiliser 143
non-communicable diseases 108
non-nucleotide reverse transcriptase inhibitors 242
North American Free Trade Agreement (NAFTA) 135
Nozick's distributive justice theories 28
NRC *see* National Research Council
nuclear energy 193–7
  Chernobyl meltdown and plutonium 195
  and CO_2 emission 194
  fusion power 197
  ITER project 197
  nuclear reactors 196
  nuclear waste 195–6
  radioactive waste 195–6
nuclear power stations 196
Nuremberg Code 239–40

obesity 109, 114, 117, 140–1, 142, 213
OECD *see* Organisation for Economic Co-operation and Development
oil 171, 173–4
  see also CO_2 emission; gas; greenhouse effect
oligopolies 61–2
  see also monopolies
organic farming 139–40
Organisation for Economic Co-operation and Development (OECD) 204–5, 208–9
'original position' 32
ownership
  land ownership 123
  private ownership 40–3, 61, 70, 258, 259–60
  public ownership 258–9

patents 228
  corporations' control of patents 127
  drug patents 227–9
  gene patenting 150
  justification of human intervention incentives 166–7
  life patents 166–9
  obstruction of research and effective health care 169
  patent protection 72, 73–4
  and *Pseudomonas* (oil degradation process) 167
  see also intellectual property rights
Paxil (Aropax) 225
PBAC *see* Pharmaceutical Benefits Advisory Committee
PBS *see* Pharmaceutical Benefits Scheme
'Peak Oil' 173–4
penicillin 222
perfect competition 40–1
personal ethics 18–19
pesticides 111, 127
  food contamination 142–3
  insect resistant crops and Bt toxin 158
Pharmaceutical Benefits Advisory Committee (PBAC) 234–5
Pharmaceutical Benefits Scheme (PBS) 234, 235
Pharmaceutical Research and Manufacturers of America (PhRMA) 234
pharmaceuticals *see* drugs
plantation production 121
plutonium 195, 196

politics
    and democracy 261–2
    domination of the political process 63–4
    and drug company profits 233–4
    political function (of society) 20
pollution 111, 117
    ammonia 143–4
    carcinogenic organic compounds 173
    factory farmed animals and pollutants 126–7
    from genetic modification 153, 157
    pollution permits 179–84, 185–6
    water pollution and dense livestock 131
Ponzi 81, 92
poverty 100, 111, 112–13, 139, 149
    aid 262–4
    and China 75–7
    death of the peasantry 135–7
Prescription Drug User Fee Act 251–2
prevastin sodium 232–3
price 37–8
    agricultural price supports and production limits 122–3
    determination of price – supply and demand interaction 38–9
    drug prices 215, 234–5, 244–5
    equilibrium price 37, 39, 53
    food prices 120, 176
    market prices 89
    price elasticities of demand 74
    vulnerability of poorer populations to food price rises 135
private equity funds 83–4
private health care 198–201, 204–6, 210, 212–14, 217
    in the developing world 216–18
privatisation 42–3, 55, 72, 75, 77–8
    drugs – privatisation and intellectual property 227–30
    of health care 216–17
    private ownership 40–3, 61, 70, 258, 259–60
    problems of privatisation 66–7
    of water resources 145
production 257–9, 266
    agricultural price supports and production limits 122–3
    determination of price – supply and demand interaction 38–9
    food production level and climate change 176
    human pharmaceuticals production – GM 148
    sponge iron producers 183–4
    sweatshop production 72–3
    see also food production
productivity 90
professional ethics 18–19
property rights 88
protease inhibitors 242
psychotropics 224–6
    barbiturates 224–5
    benzodiazepines 225
    Paxil (Aropax) 225
    risks 225
public health care 107, 202, 214–15
    in Cuba 218–20
public ownership 258–9
pyrolysis 191–2

quantity theory of money 45

rainfall 175
Rawls' theory of justice 31–4
recession 87, 121
recombinant bovine growth hormone (rbGH) 165–6
reforestation 163, 172, 174, 184
regulation 52–3, 81–3, 218
    neoliberal reform of regulation 82–3
    physician self-regulation 199
'rescue packages' 87
research 227, 234, 236, 238
    animal testing – drugs 236–8
    biotechnological research 219
    corporate control of 63
    CSIRO 166
    drug development 235–6
    and drug patents 227–8
    drug R&D budgets 244
    funding of research 150
    human research ethics 239–41
    obstruction of research by patents 169
    publicly funded research 248–50
reserve currency 77
resources 58, 73, 75, 112
    China's agricultural resources 75–7
    financial liberalisation 50–2, 79
    and intellectual property rights 43
    land reform, minerals and industrial development 265–6

natural resources patents (traditional knowledge) 229
overseas control 263
private ownership of resources 40–2
and privatisation 145, 216–17
provided by FDI 50
wastage of 64–5
see also water
retributive justice 29
Retrovir (AZT) 241–2, 243
Rezulin 249–50
Ricardo's comparative advantage theory 47–9, 72–3
rights
  animal rights 16, 143–4, 237
  autonomy rights 12–13
  democratic and legal rights and autonomy 12
  duties and rights 11–13
  economic, social and cultural rights and the UDHR 13
  genetic property rights 166–9
  human rights 12–13, 16–18, 71, 203, 229–30
  intellectual property rights 43, 167, 168
  legal rights 12
  monopoly rights for brand name drugs 228
  property rights 88
  restriction of workers' rights and bargaining power 89
  welfare rights 12, 13, 17–18
  see also carbon trading
right-wing (free market) liberalism 25–7
  law and justice 28–9
  liberty versus justice 27–8
  recognition of rights 26
  and state power 26–7
risk 51, 58, 172–3
  and derivatives 83–4
  and health insurance 200
  relative risk reductions – drugs 232–3
  risk aversion 81
  risk reduction 85–7
  risks associated with psychotrophics 225
RotaShield® (rotavirus) vaccine 253

safe currencies 76
salination 128–9, 130–1, 133, 163, 176, 191
sanitation 110–11
SAPs *see* Structural Adjustment Programs
sea levels 174–5

self-regulation (physicians) 199
share markets 80, 92, 258, 261
'sickness funds' 209
silicosis 111
Smith, Adam 36–8, 61–2
social contribution 16–17
social ethics 18–19
social mobility 58, 93–4
social policy decisions 15
society
  belief systems 20–2
  ideological and political functions 20
  material function 20
SOEs *see* state owned enterprises
soil degradation 139
solar power 187–9
solar thermal electricity (STE) 188
specialisation 71–2
speculation (share purchases) 79, 83
  and derivative trading 83–4
  shift to commodities 120–1
sponge iron production 183–4
stagflation 46, 67
state owned enterprises (SOEs) 41–2
state power 26–7, 28, 31
STE *see* solar thermal electricity
stress 114–16
Structural Adjustment Programs (SAPs) 55
subsidies 137–8, 180
subsistence farming 122, 135–7
sulfanilamide 221–2
supply 38–9, 40, 62
  and ACCC 52
  and food production 118, 119–20
  supply shortages and food prices 120
surrogate end points 247
  see also clinical trials
sustainability 64, 133–4, 265
  sustainable farming techniques 125–6
  sustainable organic farming – Cuba 139–40
sweatshop production 72–3
swidden agriculture 124, 126
symbiotic interaction 126
synthetic oestrogen 222–4

takeovers 79
tariffs 46, 49, 66
  GATT 55
  tariff barriers 71
  tariffs protecting industry and agriculture 123

taxation 18, 27, 39, 46, 68–70, 76, 123, 139
　fiscal policies 43–5
　funding of Medicare program 205
　monetary policy 44–5, 46–7, 67–8
　tax cuts 69
　and wealth concentration 260–1
technology
　agricultural biotechnology 160
　innovation and sustainability 64
　labour saving technologies in farming 118
　new technology 61–2
　and productivity 90
　technology transfer – the Green Revolution 124
　wave power technologies 188
　see also genetic modification
teleological ethical theories 11
thalidomide 227
tobacco related disease/illness 105–6, 113–14
'toxic assets' 87–8
trade
　cap and trade (carbon trading) 179–86, 187
　carbon trading 179–84
　derivative trading 83–4
　fair trade movement 266
　free international trade 47–8, 71–2, 135
　FTA and drug companies 233–5
　GATT 55
　international trade 47–8, 71–2, 74–5
　trade barriers 48
　trade deficits 77
　see also free markets
trademarks *see* intellectual property rights
traditional knowledge 229
transgenes 147–8, 164
　Bt toxin 155–6, 159
　EU's moratorium 152–3
　making GM crops 154–60
　protein production 155
transnationalisation 211–12
transport 66, 133
TRIPS Agreement (*Agreement on Trade-Related Aspects of Intellectual Property Rights*) 43, 228–9, 235
　Doha Declaration 229–30, 235
tritium 194–5

UDHR *see* Universal Declaration of Human Rights

unemployment 44–6, 61, 66, 73, 77–8, 141
　and bargaining power 89
　and free market expansion 136
　and labour 91
　mass unemployment 82
　and monetary policy 67–8
　during Second World War 106–7
Universal Declaration of Human Rights (UDHR) 13, 19, 31
uranium 193, 195–6
US Patent and Trademark Office (USPTO) 167, 168
utilitarianism 14–16
　and animal testing 237
　animals 16
　conflict-of-rights-resolving role 15
　just deserts principle 16–18, 28, 29, 57–8, 88–90
　problems 14
　unfairness of utilitarianism 14–15

'veil of ignorance' 32
Viagra 230, 254

wages 68–70, 73, 78
　dividend imputation credit system 69
　executive remuneration 68–9
　wage levels 39
　Work Choices legislation 69–70
waivers 252–3
waste 64–5, 66
　and the Enviropig 166
　from mining operations 111
　nuclear waste 195–6
　radioactive waste 195–6
water 144–5
　ammonia pollution of drinking water 144
　consumption 139
　dryland salinity 128–9, 130, 162
　insecticides and water supplies 127
　irrigation 130, 131, 132, 144
　loss of access to fresh water 175–6
　melting of ice 175
　rainfall and flooding 175
　safe water and sanitation 110–11
　sea level rise predictions 174–5
　tritiated water 194–5
　underground water reserves 132
　water dependence 139
　water pollution and dense livestock 131
　waterlogging and mallee planting 191

wealth 55, 88, 260–1
  control of 23
  distribution of 15–16, 25, 101, 102
  private wealth and democracy 62–3
  and taxation 260–1
'weapons grade' plutonium 196
welfare 68–70
  public expenditure on welfare 208
  welfare rights 12, 13, 17–18
White Paper (carbon trading) 186
WHO *see* World Health Organization
wind 187–9
women
  crimes against 104–5
  genital mutilation 104–5
  health inequalities 104–6
  maternal mortality 104
  tobacco related disease and death 105–6
  vulnerability to HIV 105
worker cooperatives 259
World Bank 42, 50, 54–5, 70, 71, 212
  and aid 262–3
  and carbon emissions 172
  classification of national economies 97
  and currency convertibility 75
  and devaluation 75
  encouragement of private health system 210
  encouragement of transnationalisation 212
  International Finance Corporation 212
  policies 100
  role in food production 134–5
  sea level rise predictions 175
World Development Report 1993 212
World Health Organization (WHO) 14, 105
  acute pesticide poisoning figures 111
  and Chernobyl meltdown 195
  investigations into antibiotics 226
World Trade Organization (WTO) 43, 49, 54–5, 71, 76, 136, 139, 228
  EU's GMO (genetically modified organisms) moratorium 152–3
  TRIPS Agreement 43, 228–9, 235